GROWING OLDER WITHOUT FEELING OLD

Professor Rudi Westendorp was trained at the Leiden University Medical Centre (LUMC) in the Netherlands, and specialised in intensive care and epidemiology. Later, he focussed on geriatrics and gerontology. In 2000, he was appointed professor of medicine, and, from 2005 to 2012, he was head of the Department of Gerontology and Geriatrics at the LUMC. In 2008, he founded and became the first director of the Leyden Academy on Vitality and Ageing, a research institute that provides training, conducts research, and initiates developments in the field of vitality and ageing. In addition, since 2012, he has been director of the VITALITY! programme, part of Medical Delta, an innovative partnership of academic and public institutions, and enterprises, in the south-west of the Netherlands. In 2015, he moved his workplace to Denmark, where he was appointed Professor of Old-Age Medicine at the University of Copenhagen.

Growing Older without Feeling Old

on vitality and ageing

Rudi Westendorp

Translated by David Shaw

SCRIBE
Melbourne • London

Scribe Publications
18–20 Edward St, Brunswick, Victoria 3056, Australia
2 John St, Clerkenwell, London, WC1N 2ES, United Kingdom

Originally published in Dutch as *Oud worden zonder het te zijn* by Atlas Contact in 2014
First published in English by Scribe in 2015

The publisher gratefully acknowledges the support of the Dutch Foundation for Literature.

Nederlands letterenfonds
dutch foundation for literature

Typeset in Kepler Std 11.5/14 pt by the publishers
Printed and bound in Australia by Griffin Press

The paper this book is printed on is certified against the Forest Stewardship Council® Standards. Griffin Press holds FSC chain of custody certification SGS-COC-005088. FSC promotes environmentally responsible, socially beneficial and economically viable management of the world's forests.

National Library of Australia Cataloguing-in-Publication data

Westendorp, Rudi, author.
Growing Older Without Feeling Old: on vitality and ageing / Rudi Westendorp.
1. Aging–Prevention. 2. Older people–Health and hygiene. 3. Longevity–Health aspects.

613.0438

9781925106916 (AU edition)
9781925228137 (UK edition)
9781925113945 (e-book)

A CiP reference for this title is available from the British Library

scribepublications.com.au
scribepublications.co.uk

Contents

AN EXPLOSION OF LIFE

In the previous century, human existence underwent a radical change. There has been an explosion of life — never before have so many people in the developed world lived for so long. It is the most drastic of the changes wrought in our society by the Industrial Revolution. Within a period of about a hundred years, average life expectancy rose from 40 to 80 years, and the likelihood of reaching the age of 65 increased three-fold, from 30 to 90 per cent. Pensioners have also made great gains; rather than ten, they can now look forward to twenty years of leisure when they retire. And then there is Madame Calment, the French lady who reached the grand old age of 122 in 1997. Babies born today can expect to live even longer; there is little doubt that some will live to be 135 years old.

All these additional years have not come to us because of a change in our bodies — whether by genetic manipulation or any other means. No, our bodies are essentially the same as they always were. Our greatly increased longevity is the emphatic result of the enormous changes we have made to our environment. Unlike before, everyone in the West now has enough to eat, we have clean drinking water available straight from the tap, and many infectious diseases have been eradicated. In addition, the chance of being killed by (military) violence has been reduced to a minimum. So it is no wonder that we no longer die in childhood, and almost everyone reaches old age. Our ability to intervene ever more effectively to counteract the effects of illness or ageing means we are living even longer.

However, our emotional and social adaptation to this revolution lags very much behind. We are truly entrenched in outdated patterns. Who brings their children up in the realistic expectation that they will reach the age of 100? What parents simply shrug off the news that their son or daughter has failed to make the grade and will have to repeat a year at school? Rather than trying to prepare their children for life in the space of just twenty years, parents today should be teaching them that learning is a lifelong process, given that they'll need to be able to cope with circumstances that are constantly changing. And what will they themselves do, once their children are grown up? The time when we lived and worked solely to provide for our children, before retiring from professional and public life, is definitively over. Now, parents of children who have flown the nest wrestle with the question of how to fill the rest of their long lives.

This is not unfamiliar to me, as a 55-year-old. Longevity is partly determined by genetics, and, with a maternal grandmother who lived to 99, I may well reach the age of 90, or even 100. Horror! What am I going to do for the next twenty years, with two grown-up daughters who get along in life excellently by themselves? Of course, I'm glad I didn't die young, and I look forward to a carefree old age. At the same time, I can see the end of my life looming ominously ahead, and I wonder if I will weather the storm well or not.

A long life is an impressive achievement, but it is also a frightening prospect. Am I doomed to spend my final years with failing eyesight and hearing, stiff and incontinent? Or are these just the normal fears of a man in his fifties, thinking things can only go downhill from here?

Not everyone sees this increased longevity in a positive light. It makes people uneasy. Some speak of a disaster that has befallen us. There are estimates that half of the over-sixty-fives that have ever lived are currently alive today. Why has no one pulled the emergency brake? The certainties of the past have given way to prospects that do not yet appear in clear focus. And this has all happened extremely quickly. When we think of getting old, many of us look to the lives of our parents or grandparents as a beacon to help us navigate the stormy seas of life. But between the time of our grandparents and the time we become grandparents ourselves there are four generations, spanning a period of a hundred years or so. That is why it is wrong to think that we can take the life stories of our parents and grandparents as a blueprint for the way our own lives should unfold. Those images are no model for the life that awaits us. We can drink deeply from their skills and knowledge, but life moves forward, not back.

If you care to listen, old people will tell you that life is hard work. Getting old is associated with loss — sometimes it is unexpected and early loss, but, increasingly, it comes slowly, and later in life. We must prepare for this and adapt accordingly. Arthur Rubenstein (1887–1982), one of the world's greatest pianists, was able to enthral audiences with his playing to a very old age. He compensated for his loss of light-fingeredness by limiting his repertoire, by practising more, and by starting pieces more slowly, so that he could more easily quicken his playing when necessary. Fortunately, older people are usually well able to adapt to this loss of their functions. The elderly are generally happy with their health. Two-thirds describe their own state of health as 'good to very good'. However, despite this positive view, many people don't want to accept the idea they, like everyone else, will inevitably get older: 'Why does that have to be so?' Funnily enough, when asked whether they want to remain healthier for longer, everyone answers in the positive. 'Of course!' is the unanimous cry. And staying healthier for longer is something we are getting increasingly better at, with the result that we are living increasingly longer lives.

Our society is similarly ambiguous about all this ageing. Students are expected to gain ever more specific skills in an ever-shorter time, or risk being too old and unsuitable for the modern labour market. But in the long run, all that haste is more likely to make them less, rather than more productive. Once you reach the age of 50 these days, a change of job is out of the question. And if you lose your job at that age, it is pretty damn hard to get another one. There are practically none to be had. Employers believe in the myth that the physical and mental abilities of 'seniors' are limited and liable to deteriorate rapidly, so investing in their bodies and brains is a waste. While some entrepreneurs recognise the untapped potential of the 'silver economy' — never before have so many older workers and consumers contributed to our economy — at the same time, ageing is seen as the cause of the socio-economic problems we currently face.

In 100 years, we have outgrown the current biological and social order, and that order is due for an overhaul.

We need to rethink our lives and adapt them to the circumstances we live in today. Developed countries have now, reluctantly, made a start by increasing the pensionable age of 65 by a few years. Perhaps it would be better to simply abolish the fixed retirement age altogether. It forces us into an increasingly ill-fitting straightjacket. Now that we are living longer, healthier lives than ever before, we have the opportunity and responsibility to fill them with meaning ourselves.

This book is a 'satnav' to help find your way through the life that lies ahead of you. In it, I show how and why human beings have adapted to their surroundings over millions of years. I also tell how our lives have constantly improved — so much so that the structure of our population no longer resembles a pyramid, but a skyscraper. Then, of course, there is the question of what we should do with this long life, now we have been saddled with it. Can we steer it in a particular direction? Everyone says ageing is 'normal', that it is 'usual', but is that really so? What can we learn from people who live on healthily into extreme old age? Does it help to eat less, or to take hormones, vitamins, minerals? What can we learn from old people who remain full of vitality, despite illness, and infirmity? How do they retain their sense of wellbeing? Of course, I also discuss extensively the social and political implications of this explosion of life.

I examine all of this through a medical, biological, and evolutionary lens. As this book intends to show, this is the only way of looking at it that can explain why we age. In this evolutionary context, where sex and reproduction play a central part, it becomes clear that the development of newborn babies and the ageing of adults are two sides of the same coin. We are genetically programmed to love our children, and our bodies are predestined to 'lapse' in old age.

In this book I attempt to give an accessible insight into the ageing process, intrinsically complex and many-sided though it is. This is also the reason I begin each chapter with a brief summary of its contents. I often start with a Dutch anecdote or a report on research that has been conducted in the Netherlands, but these findings are applicable to most of the developed world. And for those who want to know more than the chapter provides, at the end of the book I've listed suggestions for further reading.

I cannot always avoid using generalised terms such as 'youth', 'adults', and 'older people'. 'Older people' are adults for whom the ageing process has begun to take effect; this happens, on average, some time after the age of 50. I use the term 'old age' to refer to those aged 75 and over, and I describe as 'very old people' those who have crossed the 85-year threshold. At present, that is the median age at which most people die in Western countries.

Older people today are in no doubt that their lives are worth living, on into very old age. They can teach us how we can grow older without feeling old. I am very pleased to let them have their say in the final chapters of this book.

1

THE RHYTHM OF LIFE

Everything that exists, ages. That's true of books, beer glasses, washing machines, and people. Ageing, as a process, is the result of an accumulation of minute amounts of damage that we incur simply by being. From a biological and evolutionary point of view, there is no reason for us to get old. It is our early lives that count: when we have the capacity to conceive, bear, and raise children. When that is done, our useful life is over, in the biological sense. Our chronological age, health, and social standing are bound up together by the ageing process. Developments in technology and medicine mean that our biological and chronological age are becoming increasingly disconnected from each other.

Some time ago, I was invited to speak about ageing at an event organised by a debating society that meets once a year to discuss a matter of importance. The club went back a long way, and its members were also no longer in the bloom of youth, so to speak, so the subject was close to their hearts. They wanted me to shed some light on it for them.

I was given free rein to give my talk in any way I pleased; but, free or not, it is never an easy task to explain the central concepts of the ageing process to a general audience. I myself only really gained a clear idea of them a few years ago. I wanted somehow to show those people in the audience, by means of an example taken from their own experience, that ageing is a universal phenomenon — not only among humans, but all material, living or not. That would make for an interesting discussion. Perusing my bookshelves, I found an old copy of the Bible — leather bound, published in 1856. Bookworms had left it riddled with holes. I took the volume from the shelf and opened it. Several pages fell out of it as I turned them. No matter how careful I was, the leaves turned to dust beneath my fingers. The book had aged; it was now old, very old.

The story of the disintegrating Bible left a great impression on the members of that society. Just as the book had become fragile, they could feel in their very bones that their bodies had become frail. It was a revelation when they began to appreciate that *everything* ages. Then I pointed out that there is one part of the Bible that does not age: its contents. They are read, recited, sung, and reprinted over and over again. The text remains as alive as it ever was.

AN ACCUMULATION OF DAMAGE

It might seem glib to equate old people with old books, but the comparison is a valid one. I first came upon it after learning about the work of the Nobel Prize laureate Peter Medawar. It was then, in Manchester, England in 1998, that I first began to wonder why we age.

My generation of doctors, who studied in the nineteen-seventies and eighties, were only taught about the difference between children and adults. No further variety in human age was recognised by science back then. Ageing received little more than a passing mention. It was not until much later, when I was in Manchester, that I learned about the principles behind the ageing process, and began to understand them. Interestingly, to do so I had to abandon the mantle of the medical scientist, and to assume that of the biologist. Biologists seek to explain the diversity among different species and their different lifecycles — development, reproduction, and eventual demise — under very diverse circumstances. For this reason, biologists spend a lot of time thinking about ageing, and the field boasts a whole host of great thinkers about that issue, beginning with Charles Darwin. It is quite remarkable that so little of this rich trove of biological knowledge and thought has found its way into medical science; all the more so because the majority of those seeking the help of doctors these days are old people.

In Manchester, far from any patient's bedside, and in the midst of a biological research group working with worms and flies, I had the time to think about the ageing process. Why do functions inevitably decline? One of the classic thinkers I met, on paper, at least, was Peter Medawar. He received a Nobel Prize for his work in a completely different field — that of kidney transplantation. At the opening of his laboratory in 1951, he made a speech entitled 'An Unsolved Problem of Biology', an initial attempt to understand why we age. Even today, it is a pleasure to read that speech. At Leiden, we use his work to introduce medical students to our 'Ageing' course.

Medawar's text is striking in its use of plain words to explain the fact that ageing is a ubiquitous process. His comparison between glasses and human beings completely changed my perspective. When a bartender goes to pull a pint of beer but accidentally knocks the glass against the bar as he reaches for the tap, the glass will sometimes break. It may be a new glass that shatters. In that case, there must have been some tension in it due to an irregularity that arose during the moulding or blowing process — a manufacturing fault. But the vast majority of glasses that get broken in this way are simply old, and are liable to shatter at the slightest impact. A 'young' glass reacts with a resounding ring when it is knocked against the bar: the shock is absorbed by the material it is made of. But it is a very different story with an old glass, which reacts with a dull crack, and breaks. We call this 'material fatigue', an accumulation of tiny amounts of structural damage. An old glass may still look fine, but will break on the slightest of impacts — just like an old rubber band that snaps when stretched.

As with books and glasses, washing machines also break down with time — not because of any manufacturing defect, but due to wear and tear, the damage suffered by the machine over the years, which makes it liable to break even at normal, moderate levels of use. Such an appliance is simply old, and is usually replaced.

Because manufacturers know the rate at which their products age, they can estimate precisely how long the machine will last. The main factor influencing the appliance's useful life expectancy is the choice of materials and technology used to make it. So there is little element of chance involved when the machine 'suddenly gives up the ghost' and 'falls apart', requiring the purchase of a new one. After all, they're deliberately designed to do that. You can even find tables on the Internet listing how many washing cycles that various models are designed to survive.

Realising that all lifeless matter ages is an essential step toward understanding the human ageing process. We do not age by living, but simply by 'being'. This is a universal principle. Books, beer glasses, rubber bands, and washing machines age, even if they are never used. As time passes, damage accumulates in the materials they are made of, making them liable to fall apart under the slightest pressure. This is no different from the damage that occurs to the tissues that living organisms are made of, causing people to become ill and frail, and to eventually die. A general definition of ageing could be that, over the years, something or someone becomes increasingly fragile or delicate, and breaks or dies when exposed to even slight stress. This leads directly to the negative connotations many associate with ageing: 'See! It's all downhill from here.'

ALL FOR THE NEXT GENERATION

Once we turn 50, it becomes inescapably clear: our bodies demand more attention. Our bodies used to be able to get back on track easily after a bout of exertion, but when we reach that age, a day working in the garden can take some recovering from. The next morning, we find our arms, back, and legs sending us some very unmistakeable messages. Our bodies need time and rest to repair the damage done to our muscles and joints, possibly even with the help of massage or medication. The first time it happens, we suffer this discomfort with a certain equanimity. We reason that anyone who does a bit of sport or gardening can sustain an injury: 'In retrospect, maybe it was silly to try and dig the whole garden in one day.' When this pattern repeats itself a few times, we start to wonder if we shouldn't do something to get into better physical shape. How difficult can it be, after all? Fired up with enthusiasm, we head for the gym. Often, the problem turns out to be more stubborn than we expected: 'At first, training got me out of breath pretty quickly. That improved within a couple of weeks. But my muscles ache for ages afterwards, I tell you! The pain lasts a lot longer than I remember it used to. I need more time to recover, to get back on an even keel. Then I'm okay for a while, but my knee is still my "Achilles heel". It's stiff and painful, and it doesn't get better, no matter how much I exercise.' After 50 years of being lived in, our bodies seem to have become fragile — a persistent injury develops, and a visit to the physiotherapist is unavoidable. 'I never had trouble with my knee before,' we tell the therapist in consternation. Our training plan, so valiantly begun, now has to be scaled back.

Charles Darwin and Peter Medawar understood: from an evolutionary biology perspective, there is no reason at all for us to get old. The development of individuals from birth to sexual maturity is key. Humans need to reproduce and care for their children, so that they, in turn, can reach sexual maturity and produce offspring of their own. Our DNA, the blueprint for our bodies, creates an 'eternal' cycle. The term Darwin coined to describe the capacity for such cyclical repetition was *fitness.* For Darwin, this did not mean physical strength or resistance to disease; fitness is the capacity and the drive to have children — the more, the better. It is not only our physical constitution that is important in this 'fitness programme', but also our psychological makeup. And, of course, we put our own offspring first, and we feel a strong sense of responsibility towards them, since our children are dependent on us for a long time before they are ready to leave us and start their own families.

We learn to walk, talk, survive, and love — all biological functions, in fact — for the sake of this fitness programme. We are the product of natural selection, the mechanism behind evolution. Individual members of a species that are well adapted to the environment they live in have a greater chance of surviving and being able to care for their offspring than those that are less well adapted. Since those better-adapted individuals pass on the necessary traits to their offspring, well-adapted individuals will gradually come to predominate in the population. This is what Darwin called *survival of the fittest.*

In humans, this fitness programme lasts for about 50 years. The necessary information is stored in our DNA. First of all, there is the biological miracle that a defenceless little baby can develop into a unique person in the space of fifteen to twenty years. This development is rigorously programmed, and evolutionary biology helps us explain the behaviour of adolescents. Natural selection makes young adults ambitious risk-takers, thirsty for knowledge, with a craving for affection and sex. Without these characteristics there would be no fitness, and our species would soon die out.

This developmental stage is followed by adulthood. Our fitness programme makes us strong so that we can survive long enough to bring up our children. This is why vigour, decisiveness, and problem-solving abilities are also subject to natural selection. Fitness requires an optimistic outlook on life, combined with a realistic appreciation of our own abilities. In adulthood, it is also important that our bodies and minds develop in tandem with one another. Fertility, and in particular the female cycle, is complex and vulnerable; even the slightest physical or emotional stress can throw it out of kilter, with infertility as a result. Next, people have to want to have sex, otherwise no new life will be created.

And then, after a period of two generations, everything in life deteriorates. Like Buddhist monks who have spent weeks creating a mandala out of tiny grains of sand, with meticulous attention and accompanying rituals, only to brush it away in an instant with a sweep of their hand: the ceremony is over, the mandala has served its function. Many older people are able to joke about the fact that their body has started to sag, but many younger people begin to panic at the mere thought of it. Some dread the advent of their 30th birthday: 'Will I still look good?' Some men go grey or bald even before they reach the age of 30. Baldness is caused by a deterioration of the hair follicles. Greying is due to a similar process: the follicle is still healthy, but the melanin-producing cells deteriorate and are no longer able to create the pigment that colours our hair —

just like an empty ink cartridge in a printer. Irrespective of how seriously you are personally affected by baldness before the age of 30 or grey hair before the age of 40, the evolutionary fitness programme remains unaffected. You fall in love before you're 20, leaving ample time to have children before you reach 30. Baldness and grey hair do not appear before you have children to care for; by then, there is no longer any need for you to be attractive.

However, ageing is more than just turning bald and grey. The time it takes a person to complete a marathon gives a precise indication of the state of development or decline their body is in. Before you can complete a marathon for the first time, you need to have reached adulthood and to have run endless kilometres in training. If you then continue training, the time you need to reach the finish line will fall rapidly. The fastest finishing times are reached at about the age of 30. Your chances of becoming an Olympic champion later in life than that are negligible, no matter how hard you train. This 42-kilometre stress test is a ruthless indicator of the fact that your physical performance level is in decline — much earlier than many people expect. The fate of speed skaters and racing cyclists is no different. How many thirty-somethings do we see on the medalists' podium? But top physical performances are not necessary in the daily life of an average adult in their thirties, consisting of caring for home and family. If you perform physically at an average level, it will be twenty more years before a day's digging in the garden leaves you unable to get out of bed the next morning.

In old age, running speed is a good gauge of the amount of permanent damage that has accumulated in our body, and of the remaining functions it possesses. Some older people stay nimble far into old age. On average, such people live longer than their stiffer, less agile peers who have more difficulty moving. Of all the measures available to doctors to help them assess how frail and how likely to die their patients are, running speed appears to be one of the best. It indicates how well not only the muscles and joints are functioning, but also the nerves, heart, and lungs. Having reached old age without loss of mobility is a sign that the ageing process has not yet seriously affected the body.

The development and ageing of our brains follows the same pattern as the rest of our body. For most of us, it is not easy to accept that our body's functions will decline, but we find it especially disturbing when our brains begin to let us down early in life. It is common knowledge that no mum or dad, grandma, or grandad stands a chance of winning in a game of Pelmanism against their children or grandchildren, even when sheer frustration after a few lost rounds prompts them to try their hardest to win. As their tension level rises, their performance just keeps getting worse.

Children's ability to recognise visual images, to associate them with a particular time and place, and to store that information in their memory and to recall it on demand is truly phenomenal. Early in life, the ability to recognise your father and mother instantaneously, at any time, from among thousands of grown-ups, is vital — natural selection has provided our children with this remarkable skill. As we get older and more independent, we quickly become less skilled at this trick — but, still, anyone can continue to enjoy a game of Pelmanism right up to their 100th birthday and beyond. On average, enough brain function remains as residual capacity to give someone a fair game.

Maths professors are usually appointed around the age of 30. Einstein produced his most important work before the age of 40. Apparently, this is when human beings reach the summit of their theoretical and algebraic abilities. This is early, but it is significantly later than puberty, the onset of adulthood, from an evolutionary point of view. Behavioural imprinting, spatial awareness, memory, and reproductive ability are at an optimum when we are young; but, despite the deterioration of these individual cognitive abilities, humans become better at solving complex problems as adults. The effective sum of these individual functions increases because they are increasingly attuned to one another. This is also true of difficult emotional and social problems, which require empathy and managerial skills to solve, and those competencies usually come with practice. That is why the environment in which we grow up is so important. It is principally a *biological* fact that you can become a father or mother at the age of 20, or that you can be appointed to a maths professorship at 30, but it is experience and culture that turn you into an effective manager, a *grande dame*, or a wise man. This is why we do not begin to mature until our forties, and that is when we face the greatest challenges in our personal and professional lives.

RITES OF PASSAGE

For as long as humans have existed, we have marked important milestones in our lives — birth, coming of age, marriage, death — with rituals. These rituals help individuals and society as a whole to move on from an old role and assume a new one. Baptism, confirmation, our first day at school, college hazing, or initiation ceremonies are all examples of rituals used in Western society to mark times of transition. The French anthropologist Arnold van Gennep (1873–1957) called these rituals *rites de passage*. He saw these 'rites of passage' as part of the process of socialisation — the adjustment individuals make to the society they live in.

From the sixteenth century, a popular motif in engravings was the 'Steps of Life', or the 'Ages of Man', where the stages of a man's life were represented on a rising and descending staircase, charting the chronological ascent from birth to a climax at the prime of life, and then descent towards eventual death. Most engravings show the trajectory of life from age 0 to age 100, with each step on the stairs representing ten years. It starts with a baby, full of future promise, on the left, and it ends with a hunched old man on the right, a shadow of his former, vital self. Youth and old age are shown as two extremes, on the lowest steps of the staircase, as if to say that the beginning and the end of life are fixed. Depicted right in the middle, on the top step, is a man aged 50, in all his midlife splendour. But this is also where engravings differ greatly: some of these men in their prime are represented as honoured military heroes, others as successful merchants, and yet others as aristocratic noblemen. Finally, the foreground nearly always shows a depiction of the Last Judgement. The concept of the course of human life was unquestioningly bound up with prevailing Christian morality. That was undoubtedly part of the reason why prints of 'The Ages of Man' adorned the walls of so many European parlours right up until the early twentieth century.

While the majority of these engravings showed a man in his prime at the top of the staircase, some showed the 'Ages of Woman' instead. The women depicted in this way are shown reaching the top step at the age of 30! Are we then to assume that the ageing process sets in earlier for women? There is no biological reason to believe so. In fact, the opposite is the case. On average, women live longer than men, under almost all circumstances. In short, the early decline of women as shown in these engravings is primarily a reflection of their social status. These prints are a blunt reminder of how people used to think about the social progression of men and women through their lives. Worse still, some post-menopausal women were thought to possess supernatural powers. They could 'make the grass wither and the fruit shrivel on the tree'. This did not, of course, apply to high-born women.

The rites of passage depicted in the 'Ages of Man' show how closely age, health, social status, and environment are bound up together. It is for this reason that it is important to differentiate between the various manifestations of age and ageing: our chronological age, our biological age, and our social age. The relations between these types of age depend greatly on the time we live in. To some extent, age and environment are inextricably linked, but more often, this relationship is the result of the choices we make, both consciously and unconsciously.

Chronological age is the most direct manifestation of ageing — indicated by the celebration of our birthday each year. Sometimes they are special birthdays, marking particular milestones in life. This is what is called our social age. For example, we attain the status of legal majority on our 18th birthday, along with the right to vote and to marry without our parents' consent. It is simply a societal choice to designate 18 as the age of majority; not all that long ago it was still 21. At the other end of the spectrum of life, 65 is the age in many countries at which employees can legally be dismissed without further reason. This right of employers came into fashion in the late-nineteenth century, after the German chancellor Otto von Bismarck's grand gesture of paying out pensions to the very few civil servants who made it to 65 and beyond. For many, both employers and employees, this cemented 65 as the normal age at which to leave the world of work behind and to start enjoying retirement. But there are great differences between countries. The French reach retirement age at 60. In China, it is 55 for men, and 50 for female workers.

It is understandable that a society needs to have clearly defined life stages in order to function, but those definitions are based on the assumption that the quality of our bodies and minds is fixed in relation to our chronological age. A human being's sequential development is fixed and unchangeable. Babies learn to walk before they learn to talk. But the speed at which they develop can differ enormously from child to child. Exceptions notwithstanding, the human body is ready for sex and reproduction by the age of 18, but the developmental level of many people's minds by that age is far from adult. The biological development of the brain is still ongoing, and will not even begin to reach completion until we turn 30. For many, the rights and obligations of legal majority come far too soon, but for others they are attained much too late. And there is even greater variation between people when we look at the deterioration of their physical and mental state — that is, the ageing process — in later life. Some people look 80 when they are only 50; they are well past their prime. But the opposite can also be true — people in their eighties who look middle aged. This variability in physical and mental qualities is what we call our biological age, and deterioration in biological age is the second manifestation of ageing. It is not logical to send someone into retirement at a predetermined chronological age — at least, that is, if the age of retirement is to be based on medical and/or biological grounds. For some, retirement comes (much) too early; for others, much too late.

The rhythm of life — the sequence of development and ageing — does not vary between different mammal species, but the pace at which everything happens certainly does. In general, there is a relation between the speed at which a body develops and the rate at which it ages. In rodents, for example, everything happens quickly. Mice reach sexual maturity in six weeks, and rarely live longer than two years. But for other animals — humans and elephants, for example — all this takes much longer. The latter can also live to great old age; pregnancy is longer in elephants than humans, and elephant calves take a long time to develop and reach adulthood. The time required to develop by most small mammals, such as cats and dogs, is somewhere between that of mice and elephants.

Even within individual species there is a link between developmental speed and ageing. Experiments on laboratory animals have shown that rapid growth, or catch-up growth after a period of food scarcity, is associated with accelerated ageing. Studies have also been carried out to investigate whether there is a link in humans between the onset age of puberty and/or the menopause, and the development of disease in old age. These studies revealed a positive correlation between a late onset of puberty, body height, and bone strength. The conclusion is that a lengthy period of development gives a better result, biologically speaking. However, the other side of the coin is that tall people have an increased risk of developing cancer. The prevailing interpretation is that excessive growth can have negative consequences.

Unlike chronological age, which is immutable, there appears to be a great deal of possible variation in the biological development of individuals of one species, and this is what biologists call 'plasticity'. A dog is a dog and a human is a human, but each individual can have a very different life story. Nematodes — also called roundworms — can undergo a temporary metamorphosis in reaction to harsh conditions. This is the worms' so-called '*dauer*' stage (from the German word for 'endurance') in which they enter a kind of stasis, do not reproduce, and are able to resist adverse external influences. Nematodes can survive long periods under unfavourable conditions in this way. When conditions improve, the worms simply resume their normal lives and begin reproducing again. Scientists are keen to discover the principle that underlies this *dauer* stage — not only because it is enigmatic in itself, but also because it could potentially be hugely important in human medicine. Nematodes, with their ability to enter a *dauer* stage, have a lifecycle characterised by a long, disease-free existence, seemingly without having to pay a biological price for this.

Not every species is endowed with the ability to change its lifecycle so deeply and yet so flexibly, depending on the surrounding conditions, by simply 'taking a break'. Human beings' talent for this is small. Bears that hibernate have slightly more talent, but the greatest talent in this area must be that of the abovementioned nematode worms. 'Plasticity' is seemingly written into their genetic code, and for some species it is a necessary feature to protect them from extinction.

The historical period we live in has a great influence on the way ageing manifests itself. A comparison of the present with a century ago shows that the course of our lives is closely dependent on our environment. Whereas life expectancy stood at about 40 years just a few generations ago, it has now more or less doubled. And we are now staying healthy for much longer. While people in a traditional setting were fully prepared for life by their eighteenth birthday, those reaching adulthood now are expected to have many more mental and emotional skills than their predecessors, but they still have a lot to learn beyond the age of 18. And, if anything, those expectations seem to be increasing rather than decreasing.

In the modern world, we must continue to develop and adapt throughout our lives. Social and technological developments happen so quickly that acquired knowledge and skills are soon outdated, and people are soon no longer adequately equipped to function in society. This may be why men today reach the apex of their career well before they turn 50. Conversely, women are no longer sidelined from society at the age of 30. It is difficult to discern any patterns in all these changes. Each era, each society, has its own opportunities, moral code, and customs. But, of course, we cannot completely detach all the social expectations we place on ourselves or each other from our biological age, the moment we reach adulthood, and the time we become sick and dependent on others.

The good news is that the modern age offers us more possibilities than ever before. Inspired by the Paris student revolts of 1968, activist groups sprang up all over the world, which aimed to break away once and for all from traditional male and female life trajectories as depicted in those 'Ages of Man' prints. They wanted to untie the knots connecting chronological, biological, and social age, as they felt they had become obsolete. Major advances in medicine and technology added momentum to the developments taking place in society. Sex and reproduction, which until then had been directly linked to chronological and biological age, now became a matter of choice. The contraceptive pill gave both women and men the power to choose whether and when sex should lead to reproduction. With this, the concept of social age also began to run adrift. Marriage was no longer the rite of passage into parenthood. The responsibility of having children could be postponed at will, or entered into outside of wedlock. Here, too, advances in medicine, and the loosening of social constraints mutually reinforced each other. In vitro fertilisation allowed otherwise infertile couples to have children with the aid of a donor. It now also became possible for women to have children at an age that was previously impossible. In short, the desire for a new, more self-determined, and individual life trajectory has become reality because issues of fertility, ageing, and chronological age have been, in part at least, disconnected from each other.

Adapting to the environment — socialisation, as Arnold van Gennep called it — is essential for survival and reproduction to create the next generation. Natural selection plays a crucial part in this. The plasticity of the nematode worm, with its ability to change form and to switch off the ageing process in doing so, is not something we human beings share. The rhythm of human lives is largely fixed. But we are increasingly able to influence biological processes that used to be thought of as immutable. It is fascinating that we have been able to break down the rigid relation between chronological age and biological age in the past 50 years because it suits us better socially.

2

ETERNAL LIFE

Hydra is not just the name of a many-headed mythological monster. It is also the name of a very special freshwater polyp. This little animal appears to be an exception to the rule that everything ages. This is due to its extraordinary ability to repair itself after sustaining damage. Humans have only a limited capacity for regeneration and repair. Many types of damage are irreversible — like the loss of a finger, for example. However, some people have an above-average capacity for damage repair. They live longer than others and get sick less often. Although human beings will never become immortal, we can postpone the effects of ageing caused by damage.

In summer, ponds burst into life. Marsh marigolds flower, dragonflies flit over the surface of the water, and the pond edge offers nesting sites for many species of bird. A child's fishing net is enough to scoop all manner of wildlife out of the water: sticklebacks, for example, but also hydras — a kind of freshwater polyp.

The hydra belongs to the family of animals known as Cnidarians, which also includes corals, jellyfish, and sea anemones. Hydras are about half a centimetre long, with a tube-like body. In fact, they are little more than a large gut, with a head at one end and a tail at the other. At the head end, the hydra has a set of tentacles that it uses to catch its prey. It has been claimed that hydras are immortal. The animal's name refers to the many-headed monster of Greek mythology whose lair was in Lake Lerna, at the entrance to the Underworld. According to legend, the serpent could not be killed, because for every head that was cut off, it grew two more. Finally, it took the superhuman powers of the hero Hercules to slay the beast, which he did as the second of his Twelve Labours.

Like its mythical namesake, the hydra from the pond is also in fact mortal. If it is removed from the water and placed in the sun, it will dry out and die. One gulp from a stickleback, and a hydra is history. Nothing remains of the myth, but, still, there is no doubt that the hydra is an extraordinary animal.

DAMAGE AND REPAIR

In the late nineties, the biologist Dr Daniel E. Martínez decided to really put these hydras to the test. He filled an aquarium with water containing a large number of hydras, and waited to see what would happen. According to a rigid routine, the hydras were fed, the water changed, and the temperature adjusted to make sure it remained constant.

Not much happened. Clinging to the bottom of the tank with their 'feet', the polyps swayed back and forth, waving their tentacles around to catch food. Very occasionally, on inspection in the morning, a hydra appeared to have died. It was a rare occurrence, which did not increase in frequency as the years passed. Martínez spent four years observing the fortunes of his hydras. Then he decided it was time for a holiday, and left a colleague looking after his aquarium.

Returning from his holiday, Martínez found that all his hydras were dead. Had they been overfed? Or underfed? Was the water too warm, or too cold? Nobody knew.

However, the unfortunate fate of the hydras led to a major scientific observation, from which two conclusions can be drawn. First, the claim that hydras are immortal was demonstrably false, in spite of the Greek legend. Hydras do not live forever. The second conclusion that can be drawn is more interesting. If, during the four years the hydras were under observation, one occasionally died, and the number of deaths did not increase as the creatures got older, their chance of dying had to be constant over time! Their risk of death did not increase with increasing age, as it does for beer glasses, rubber bands, washing machines, and human beings. Up until Martínez's fateful holiday, his hydras were no more likely to die at the age of four years than they were at four months or at four weeks. Also, a four-year-old hydra looked exactly the same as a four-month-old and a four-week-old specimen. Based on this information, Martínez drew the remarkable conclusion that hydras do not age.

Having been persuaded that all lifeless matter and all living creatures are doomed to become increasingly fragile and delicate with increasing age, we were now faced with the fact that some organisms do seem to be able to escape the ageing process.

When Martínez's research group published a paper on their results, it caused quite a stir in the scientific community. His conclusion was described by many as implausible. His fellow researchers vigorously disputed his conclusion, arguing that there had to be an alternative explanation for his observations. One suggested explanation was that the period of study was too short, and so an increase in the risk of mortality had been overlooked. Martínez reasoned that four years — four times four seasons — was a respectable lifetime for such a little creature, and that signs of ageing would have been observed over that period if they had been there. Other organisms of comparable size and living in similar environments definitely show features of ageing. However, this comparison was apparently not convincing enough to change the minds of Martínez's critics.

In science, a lone observation is as useful as no observation at all, because, according to the traditional scientific method, results must be tested and confirmed, or else disproved, by other laboratories. As it turns out, many initial results claimed in leading scholarly journals cannot be replicated by other researchers — not because the reported results were made up and the scientists who published them deserve to be unmasked as frauds, but because unusual, one-off events can occur during any testing process. If this turns out to be the case, the results cannot be confirmed, no regularity in them can be claimed, and there is no reason to add them to the body of current scientific knowledge. Since 2005, scientists at the Max Planck Institute in Rostock, Germany, have been studying two thousand hydras in the institute's tanks. They want to find out whether there is any scientific predictability to Martínez's results. Eight years have now passed, and, while the occasional hydra dies now and then, the number of deaths has not been observed to increase as the polyps' age increased over time. And the hydras still look the same as they did at the start of the experiment!

How is it possible that a hydra in a laboratory tank can live to a great age without any permanent damage accumulating in its body, and can avoid falling prey to the ageing process? If a lab-tank-dwelling hydra is cut in two — a radical experiment of the type often carried out by mischievous boys on earthworms — both halves will develop into a complete hydra. The head grows a new tail, and the tail grows a new head. Unlike mice and men, hydras appear to have huge regenerative abilities. In other words, when damage to its body occurs that cannot be repaired, the damaged tissue is simply replaced. In a similar way, a salamander can regrow its tail after losing it in a tussle with a predator. We humans, however, who also sometimes lose an arm or a leg, have to face the rest of our lives without the missing limb. Our bodies have a 'segmental' regenerative power: certain tissues can be replaced, but others cannot. If up to a tenth of the human liver dies or is removed, the organ will grow back to its normal size. Our skin flakes off every day; luckily, there are stem cells in the deepest layers of our skin that replenish it layer by layer. The same happens in the gut, where the 'inner tube' is regularly worn away. This regenerative ability of cells, tissues, and organs stops hydras from ageing and keeps them looking 'young'. We now know that hydras have 'totipotent' stem cells with an unlimited capacity to divide and produce cells that can be used to rebuild any kind of tissue, and even an entire body.

This means that no damage, whether external or internal in cause, has lasting consequences. But if all the stem cells that make up a hydra are lost at the same time — for example, when it is removed from the water, placed in the sun, or gulped down into the stomach of a stickleback — the animal turns out to be mortal after all.

It seems almost inconceivable that a hydra can carry on regenerating for its entire life. In fact, such regeneration is the most common way these polyps reproduce. By cloning one single cell, they grow a bud on the outer wall of their bodies that develops into a new individual. When it breaks away from its parent body, the clone can live independently. We humans do not possess this ability, of course. Nor can we always completely repair damage we have sustained, as we simply do not have the necessary regenerative powers. This means such damage is permanent, and permanent damage accumulates over the years. Our bodies and our brains become frail and vulnerable. We get biologically older, while hydras remain 'forever young'.

Knowledge of the limited capacity of cells, tissues, and organs to regenerate has given scientists an insight into the direction they need to take their research if they are to find a way to slow, or even halt, the ageing process. For instance, Parkinson's disease is known to be caused by the death of a small set of highly specialised brain cells, known as the *substantia nigra.* This results in a shortage of the neurotransmitter dopamine in the brain. As a result, patients with Parkinson's are sluggish and stiff, and subject to involuntary movements such as tremors. Treatment for Parkinson's currently consists of prescribing drugs that increase the concentration of dopamine in the brain. The results of this treatment are good, but it cannot remove all the symptoms of the disease, and is associated with side effects. In fact, the drugs increase the level of dopamine in the entire brain, and not just the substantia nigra, where low levels lead to Parkinsonism. It would be wonderful if, in the future, scientists are able to grow a new set of highly specialised dopamine-producing cells from a sample of the patient's own stem cells. If such cells can then be introduced into the substantia nigra and persuaded to take up their normal function there, Parkinson's disease will become a thing of the past.

From a biomedical point of view, old age is necessarily associated with lasting damage to our body. Some damage is greater than other damage, but what is more important is that some damage can be repaired, and some cannot. A shattered bone in the lower leg — sustained perhaps in a motorbike accident — will hopefully leave no lasting effects that might be detectable in years to come. But a missing tip of the middle finger of your left hand — after a careless moment with the electric saw — will never grow back. We simply do not possess the regenerative power to replace an entire joint of our finger. We do have some medical tricks up our sleeve: when the lenses in ours eyes become so clouded by cataracts that reading is no longer possible, even with the aid of spectacles, modern ophthalmologists can replace them surgically with artificial lenses.

Just as ageing can be seen in the state of our eyes, other parts of our bodies and brains can sustain damage that can be repaired or reversed to varying degrees. Genuine flu is associated with terrible muscle pain and high fever, and the virus causes a lot of damage. It is not without reason that a bout of flu leaves us out of action for some time. After a couple of weeks, however, everything is back to normal. Not all viral infections have such a favourable outcome, and some cannot be overcome without leaving permanent damage. The measles virus, for instance, can cause lasting abnormalities in the brain, and the polio virus can lead to permanent paralysis. The damage caused by the human papilloma virus (HPV) can lead to cervical cancer in the long term. This is why we have vaccination campaigns for measles, polio, and HPV: they offer the only protection against lasting damage.

The above examples might make us think that all the damage incurred by our bodies comes from external sources — accidents or infections. This leads us to seek out ways in the world around us to help us stay healthy longer, since someone who manages to avoid any kind of stress or infection throughout her life will also suffer less damage. It is true that it pays to be cautious. But this does not mean that cautious people sustain no damage in their lifetime. Every time a joint in our body moves, two layers of cartilage rub against each other. They are extremely smooth and elastic, and they are very well lubricated, but even all this cannot completely prevent wear and tear. Knees and hips are particularly notorious for this, and many people suffer greatly with them in old age, because the cartilage in those joints has simply worn away. The stem cells in the deepest layers of cartilage, the source of growth and repair, are exhausted. This means that the layers of cartilage above them gradually wear away, and eventually leave bone grating against bone. Doctors call this loss of cartilage 'osteoarthritis'. It is not only knees and hips that can be affected; shoulders and even the joints in your fingers can be stricken with osteoarthritis — as can, in fact, any part of the body where cartilage is found, including the back. Wherever the health of a joint is compromised by abnormal development, growth, or an accident, cartilage is lost even more quickly, because the remaining cartilage in the joint is then additionally stressed by overuse. In this way, one kind of damage can cause or exacerbate another.

Does this mean, then, that we should always be cautious, and spend our lives just sitting still? No, because even when we are at rest, our body is in action. Although the muscles in our arms and legs might be at rest, the same is not true of our rib cage, which we use to breathe, or of our heart, which must constantly pump blood around our body. And what about the valves of the heart, which open and close with every heartbeat? Careful examination has shown that the valves of most older people's hearts are leaky and constricted. This often causes no negative symptoms; but, whether it goes unnoticed or not, the damage is there.

LONGEVITY IN FAMILIES

The idea of wear and tear is the basis for the popular but erroneous 'rate of living' theory. The reasoning is that a knee is built to bend a particular number of times and no more, and that the valves of the heart can open and close only a pre-determined number of times. When the maximum number of bends or heartbeats has been reached, it's over: the organ is broken, the body is sick, and the person dies. Proponents of this theory often point to the hummingbird as an example. It has a very rapid heartbeat and a short life. But one observation does not make a rule of biological science. A hypothesis can only be said to be substantiated if the postulated connection is observed repeatedly in different species. According to this, if the rate-of-living theory were true, it would mean athletes are busy shortening their lives by a significant margin as they train. Every time they exert themselves, athletes raise their heart rate to an exceptionally high level, after all. But, instead of shortening their lives, the repeated exercise that athletes take improves their physical fitness, and, for most of them, this results in a longer, not a shorter, life.

From a medical and biological point of view, it is also hardly conceivable that the rate-of-living theory could be correct. After all, as we go through life, our bodies do not only accumulate damage; they also have the ability to repair it. The body is more than just a passive punching bag that absorbs every blow, and slowly declines. The body can respond; it tries as best it can to repair any damage it sustains; or, where possible, to replace tissue without leaving lasting damage or scars. All the indications are that regular physical exertion stimulates our body's abilities to repair itself, so that, despite the increased physical stress the body is exposed to, the balance between damage and repair is positive, and lasting damage is less likely to occur.

Hydras appear to be able to repair all the damage their bodies sustain, but that may be because a hydra's body is not very complex. Humans' ability to repair tissue and organs is very different. It is logical to conclude that a great ability to repair cells, tissues, and organs contributes to longevity, and that the capacity for repair is contained in our DNA, our genes. This means that it is 'embedded' in us, and is one of the things that influences how long we live, on average. A greater capacity for repair would certainly extend our lifetime, giving us more years to experience the journey of life, but this is something that is not granted to us as a species. However, we have no reason to lament this state of affairs. Our bodies do have a great capacity for repair, and that is one of the reasons why we live considerably longer than mice, cats, and dogs do.

In some families, thrombosis, cancer, or mental illness occur more often than the average. Such families are genetically predisposed to this, since the structure of their DNA means these conditions are more likely to develop, and are more likely to appear earlier in life. This can be compared to a congenital defect. But the opposite also occurs: members of some families not only live to an extraordinarily great age, but they also develop few or no illnesses. One explanation for this might be that they have lived extremely careful lives, and so managed to avoid any serious accidents or infections. But it may also be the case that members of such families are genetically equipped with an above-average capacity for repair, and so incur less lasting damage, fall ill less in later life, and remain healthy for longer.

The idea that some families have an above-average capacity for damage repair is currently being tested in research. Scientists in Italy and America are studying the 60-to-80-year-old children of centenarians, and comparing these 'children' to their peers from the general population. This research aims to investigate the differences between two groups of adults that could explain above-average longevity.

But what if the fathers or mothers of these people were just lucky, and none of their aunts and uncles lived to great old age? There would be nothing special about that. In order to be certain that longevity has a biological basis, professor of molecular epidemiology Eline Slagboom and I took a different approach. We identified families as exceptional only if a nonagenarian had at least one living sibling who was also over 90. Looking at a photo of a whole set of such older brothers and sisters, anyone would exclaim, 'This can't be a coincidence — there must be something special about that family.' And it is this 'something special' that we are searching for.

As part of the Leiden Longevity Study, some four hundred of these particularly long-lived families were monitored between 2002 and 2005. It was not only the nonagenarians who were asked to take part in the study, but also their grown-up children and their partners. We drew up their pedigrees and compared mortality in these families with that of the general population. An analysis of these figures showed that every generation of the families under investigation had a 30 per cent lower chance of dying at any given age than the average in the Netherlands. That translates into an extra six years of life expectancy. We also investigated members who had joined the family through marriage — the 'in-laws', so to speak. As we expected, these relatives by marriage appeared to have no survival advantage; they reached the same age as the average Dutch person. This strict division along bloodlines in the ability to live longer indicates that there is a genetic predisposition for longevity.

The 50-to-70-year-old children who took part in our study were also compared to their partners, with whom they had weathered the ups and downs of life for many years. A similar comparison was not possible with their nonagenarian parents, since most of them had already outlived their partners. At first glance, there appeared to be little difference between the 'children' and their partners; similar age, similar size, similar weight. Also, their lifestyles did not appear to differ too much from each other: a little smoking, a little sport, a little drinking. Seen through a biomedical lens, however, there were clear differences. The children of long-lived parents had a significantly lower incidence of high blood pressure, diabetes, and cardiovascular disease. Blood tests also showed that they had a much better risk profile, with lower levels of 'bad' cholesterol and blood sugar.

The results from the Leiden Longevity Study are broadly consistent with the findings of those Italian and American researchers who also investigated the children of centenarians. They also found less cardiovascular disease, and lower risk factors such as high blood pressure, diabetes, and high cholesterol levels, among those offspring. The preliminary conclusion from our research is that families that live longer on average have a more efficient metabolism — the body's energy supply and fuel-combustion system — which means that less damage can occur to their blood vessels and organs.

Thus, damage can be prevented by having a well-functioning metabolism. But, of course, the repair and replacement of cells or tissue when they do get damaged also helps. Some scientists have made it their mission to investigate this aspect, reasoning that damage can never be completely avoided. They have set themselves the goal of finding out whether people born into long-lived families have not only a better metabolism, but also a greater capacity for repair, making them less susceptible to sickness. In short, they want to know whether long-lived individuals are in some ways similar to hydras. And the next question will then be: how can we make use of this knowledge to help us stay healthy for longer?

3

WHY WE AGE

Ageing is not a mechanism to prevent overpopulation. The history of Ireland shows that this is simply a common misconception. Ageing is a side effect of the evolutionary programme for life, whose goal is for our DNA to be passed on to our children. When we are 50, and our children are grown up and able to have children of their own, our work is done. Our body is allowed to age and can be 'thrown away'. Known as the 'disposable soma theory', this idea is based on the assumption that a person's development follows economic principles. When resources are scarce, two options are available: to invest in one's children or to invest in one's own body. Since each can only be done at the expense of the other, having lots of children means reducing one's own life expectancy. This has been demonstrated in experiments with fruit flies, and in research among the British nobility.

By 2015, there were over 7 billion people living on planet Earth, and the population was rising by about 200,000 a day. Many people are extremely worried that, if this number continues to increase, a time will come when we can no longer produce enough food to ensure that everyone on the planet will have at least enough to eat to survive. Unbridled population growth could jeopardise the future existence of the human species. How can we keep the size of the population in check? We often hear the reasoning that such a thing as ageing exists so that people will die in time, thus preventing the population from growing too quickly. The ageing mechanism is said to be determined by our genetic makeup for that very purpose, thereby preventing our species from dying out.

Of course, it is not only humans who face the risk of overpopulation and extinction, but all species. Apart from ageing, other mechanisms are said to exist to limit the number of individuals of a species. Take, for example, an island where the population of lemmings has grown too large, and food shortages threaten to cause massive mortality. The mass suicide of some of the lemmings, by leaping off a cliff into the sea, is then said to prevent a species-threatening shortage of food. This is often quoted as a further example of a genetically programmed mechanism to limit the population of individuals within a species. Although such selfless behaviour on the part of lemmings appears logical, these animals do not, in reality, sacrifice themselves for the general good. The 1958 Disney documentary film, *White Wilderness*, showing this behaviour appears to have been staged. Similarly, the reasoning that there is a genetically determined ageing programme in order to curb the growth of human populations is also incorrect. In the following section I will use the example of the history of Ireland to show that the relationships between overpopulation and scarcity of resources, self-sacrifice, and ageing should be interpreted in an entirely different way.

AGEING IS NOT NECESSARY

Thomas Malthus (1766–1884) was a British demographer, economist, and clergyman. He is best known for his gloomy theories of demographics — the study of the size and structure of populations. In 1798, he published a pamphlet entitled *An Essay on the Principle of Population* in which he predicted that the global population would increase to such an extent that food production would no longer be able to keep up. The tipping point at which there would be a shortage of food is known as the 'Malthusian ceiling': the maximum possible size of the population in relation to the yield from available land. Once the ceiling was reached, famine would inevitably occur — the 'Malthusian catastrophe'. Mass mortality would restore the balance between the human population and agricultural production.

Malthus's work was extremely influential at the time it was published. Charles Darwin wrote that the essay was crucial to the development of his theory of natural selection.

As an economist, Malthus was witness to the start of the Industrial Revolution in England. This must have been a thrilling time, when entrepreneurs were able to realise large-scale projects that would previously have been considered impossible. But, as a man of the church,

Malthus must also have been struck by the human misery that this process of industrialisation brought with it. The swift pace of industrial development created a huge demand for labour, prompting a massive wave of migration from the countryside to the towns. The number of city-dwellers increased rapidly, as did the inhumane conditions resulting from this swift rise in the urban population. Child labour, poverty, and alcohol abuse were rife. In addition, the depopulation of rural areas caused a severe drop in agricultural production, and it became necessary to import food from Ireland to provide 'fuel' for the English cities, the engines of economic development. But the population was growing at an explosive rate in Ireland, too, and agricultural yields were too small at that time to feed both the Irish and English populations. These developments led Malthus to reason that sooner or later everyone on Earth would be faced with great poverty and famine. For this reason, he called for an interventionist population policy to limit the number of children born, beginning with the poor. After all, they were the ones who had insufficient means to bring up their offspring and secure their future.

Even during his own lifetime, Malthus saw that his predictions were not always true. The potato had been imported to Europe from South America by Spanish explorers. It proved to be an extraordinarily good source of food — rich in vitamins, and high in calories. But the potato was not cultivated on any great scale, perhaps because the plant's stalks and berries are poisonous. This changed around the year 1800, when potatoes began to be cultivated on a massive scale in Ireland. The success of this newly introduced crop staved off the Malthusian catastrophe.

Despite the non-appearance of the catastrophe in Ireland, it remains the case that there is indeed a relationship between environmental conditions, the amount of food available, mortality, and population growth. Malthus was not completely wrong. Ten years after his death, in the period from 1845 to 1849, almost all potato crops failed due to the 'blight', a fungal infection whose spread was accelerated by the practice of monoculture and the damp climate in Ireland. Once more, food shortages occurred, and this time they did indeed lead to great starvation and death. During the Irish Famine (1845–1850), an estimated one million Irish people died, and a similar number emigrated. The population fell by a quarter. Malthus turned out to have been right in predicting that the 'land' — the amount of food produced — can impose a check on population size. However, the Irish population was ravished not only by hunger, but also by disease. At that time, a person's chance of dying from infectious diseases was large, especially among the poor, who were not well nourished. Typhus, cholera, and many other epidemics caused huge waves of deaths, and the young were hit hardest. Poverty, starvation, and infectious diseases formed an inextricable trinity of misery that claimed a large number of human lives.

The assumption that there is a mechanism for ageing stored in our genetic code in order to check population growth is not warranted. On the contrary, when we see that harsh conditions have caused mass mortality many times among the hundreds of thousands of generations that have reproduced, we recognise that no such genetic mechanism has been necessary to keep the number of people on the planet in check. And the same reasoning applies to all species in the animal and plant worlds.

The question should really be posed the other way round: how have humans managed to avoid extinction throughout the millions of years of their evolution? A simple answer is: because we invest as much as we possibly can in the next generation. If we were ever unable to keep enough offspring alive, especially when the circumstances around us (suddenly) become unfavourable, then we humans, or any other species in the same situation, would rapidly disappear from the face of the Earth. The past, both recent and distant, abounds with examples. Dinosaurs died out 65 million years ago as a result of climate change. Around the year 900 AD, the Maya culture in Mexico almost completely disappeared, probably because of prolonged drought. In today's world, we interfere with the habitats of plants and animals, and then worry about the decline in biodiversity.

How can we explain the fact that individual members of a species continue to invest adequately in their offspring? In his 1876 autobiography, Charles Darwin wrote:

> In October 1838, that is, fifteen months after I had begun my systematic enquiry, I happened to read for amusement *Malthus on Population*, and being well prepared to appreciate the struggle for existence which everywhere goes on from long-continued observation of the habits of animals and plants, it at once struck me that under these circumstances favourable variations would tend to be preserved, and unfavourable ones to be destroyed. The results of this would be the formation of a new species. Here, then, I had at last got a theory by which to work.

This oft-quoted passage shows how important Malthus's ideas were for Darwin's theory of natural selection. Darwin noted that humans and animals are able to produce far more offspring than the resources in the environment can support. Their reproductive programme is geared solely towards producing offspring. At the same time, an overabundance of offspring leads to competition between those children. Since genetic variation means that brothers and sisters differ slightly from one another, some siblings have a better chance of survival than others. These survivors are the ones who will pass on the reproductive programme to their offspring. In this way, the mechanism for reproduction and survival persists through the generations.

Around the year 1800, when the environmental conditions were relatively favourable, the average life expectancy in Ireland fluctuated around the 40-year mark. That is a minimum life-expectancy value for the human species, because if it sinks to below 40, the population will begin to shrink. Put more precisely: the chance of dying is then so high that a couple has insufficient time to raise at least two children to adulthood. During the time of the Irish crop failures, life expectancy dropped well below 40 years. More than half of newborns failed to reach the age of five, killed by a combination of cold, infectious disease, and famine. But even if they survived childhood, people in Ireland were not guaranteed a long life. Only the nobility could escape the bleak Malthusian scenario. They always had enough food and shelter; they could flee epidemics of infectious diseases, and that meant they lived longer. Their average life expectancy stood at around 60 years, even at that time.

Thanks to technological innovation and the mechanisation of agriculture, food production has increased continuously — something Malthus could never have imagined possible — and more people today are living longer lives than ever before. Also, the interventionist population policy expounded by Malthus to keep the number of children in check has had little effect in practice, insofar as it has been put into practice at all. The number of people living on the planet has continued to rise. The reason for this is not only that children are still being born in great numbers and staying alive, but also that we are living ever-longer lives. If, as is the case, the number of people on the Earth is continuing to rise, we must conclude that the purpose of the ageing process is not to keep population growth in check.

THE 'DISPOSABLE SOMA'

If ageing is not a mechanism to prevent overpopulation, and if some species, such as hydras, do not age, why must we 'suffer' the ageing process? Why do our bodies and our minds become infirm, and why must these infirmities, as well as illness, eventually kill us?

There is one part of us that does not seem to age: our DNA, which contains the genetic code of our lives. Copies of our DNA are preserved in our progeny, and in this way, that code is protected from degeneration. So the question of why we age can be put more precisely: why do our bodies and minds decline as the years go by, while our genetic code remains intact?

The English gerontologist Tom Kirkwood came up with an answer to this question in the form of his 'disposable soma theory' (*soma* is the Greek word for 'body'). Individual members of a species, including humans, are selected for fertility — the ability to produce offspring. Those offspring will carry a copy of the parent individual's DNA. Obviously, those individuals who die young will not contribute to the next generation. The same is true of infertile individuals, irrespective of how long they live. So, when a childless couple dies, a branch of the family dies with them. In England, for instance, childlessness was common among the nobility. Over the centuries, much has been written about 'extinct peerages' — noble family lines that have died out. Through natural selection, species — including humans — have adapted their way of living to their environment to allow them survive long enough to reproduce. Once its children have been born and raised to adulthood, an individual's body becomes 'disposable', since the continuation of its DNA has then been guaranteed. In other words, ageing is 'allowed' to occur. From an evolutionary point of view, after 50 years of faithful work, our bodies can be left to run down and decay: our DNA sloughs its old skin like a snake.

The disposable soma theory is based on the idea that the development, growth, and survival of individuals follows fundamental economic principles: the scarce resources that are available must be distributed optimally among several biological processes. To put it simply, there are two options: either individuals invest in fertility and reproduction, or they invest in maintaining their own body. One kind of investment always comes at the expense of the other. There is no doubt that investing in both body and mind are necessary and expedient. If a congenital abnormality, a physical defect, or risky behaviour means that an individual fails to reach adulthood and produce offspring, or if individuals fail to provide for their offspring, then, from an evolutionary point of view, all is lost. This is why humans have a body that will last about 50 years or so, depending on individual genetic predisposition, in order to engage in sex, reproduction, and care of their young.

The question is whether there is an evolutionary advantage to possessing a body and mind that remain functional for longer than 50 years. There is nothing 'wrong' with living beyond 50, unless investing in an above-average body comes at the expense of the investment necessary for reproduction. In the latter case, the individual may well live longer than average, but will produce fewer offspring. And natural selection will not allow those two characteristics to exist side by side: after all, the fitness programme that steers us is not geared towards longevity, but towards creating the next generation. The production of many offspring is what is maximised by natural selection. It is logical, but also contentious, to infer that investing more in maintaining our own bodies, allowing us to continue looking good and feeling healthy well beyond the age of 50, does not serve any evolutionary purpose. By then, the continuation of your DNA through your children is already guaranteed.

The biological ageing mechanism can be compared to the mechanics of an old jalopy juddering to a halt on the road. Timely inspections of the engine by a mechanic, regular tuning and adjustments, and replacement of damaged parts could have prevented the breakdown. But preventive maintenance is costly; the necessary time and money are not always available, and so it is often carried out inadequately. Our bodies, too, require preventive maintenance, and that is a complex business. In every cell of the body, damage to the DNA must continuously be identified and repaired to make sure that the genetic code it contains remains unaltered. After all, our DNA is the blueprint for the functioning of all our cells, tissues, and organs. To achieve this, an ingenious biological mechanism has developed: large knots of cooperating proteins ride like trains over the rails of the DNA molecules, identifying damage and repairing it on the spot as they go. This process requires a lot of investment. But it prevents permanent damage to a cell's DNA, which can lead the cell to produce the wrong proteins, or to go 'off the rails' and cause cancer.

The proteins that make up cells, tissue, and organs may themselves be damaged over the course of time. These proteins are complex, folded structures whose functions depend on the way they are folded. To assist in the correct folding of these proteins, cells are equipped with other, specialised proteins called 'chaperones', which initiate and guide the folding process. Sometimes, proteins spontaneously unfold, or become damaged in some other way, lose their functionality, and have to be replaced. Some proteins, such as those in the lenses of our eyes, or in our brains, are unique; they cannot be replaced, but they can be repaired. Chaperones can help some proteins to re-fold back into the correct shape. But, just like repairing damage to DNA, this process also requires a great deal of investment, and so not all proteins are repaired or replaced at any cost. The proteins in the lenses in our eyes, for example, are originally fully transparent, but the older we get, the more they congeal like those in the white of an egg, becoming increasingly opaque. The result is cataracts.

More resilient lenses are evolutionarily superfluous; humans only need keen eyesight for the duration of two generations. This is the reason people who require cataract operations are rarely younger than 50 years of age.

The disposable soma theory offers an explanation for why ageing is part of our lives, but also for why life expectancy differs so greatly from species to species. Mice develop in a very brief time into sexually mature adults, gestate their young for a short time, and give birth to a large number of offspring per litter. These investments in producing the next generation come at the expense of investing in maintaining their own bodies. Under natural circumstances in the wild, the average lifespan of a mouse is just a few months. In their natural environment, predation, cold, or lack of food means they are not blessed with a long life. That is why you seldom see an old mouse in the wild; they are all youngsters. They do age when kept as pets, or held in optimum conditions in a laboratory. When they reach old age, mice go grey, lose muscle strength and mobility, develop cancer, and die after a maximum lifespan of three years.

Changes in the environment can exert a great influence on the length of time it takes individuals to develop to sexual maturity, and on the structure and length of individuals' lives. The evidence for this abounds. When circumstances are unfavourable and the chance of dying is high, the evolutionary pressure to produce more offspring increases. More progeny are needed. There is then selection in favour of those individuals that are able to reproduce at an early age, even to the detriment of their own ability to survive for a long time. We see this in mice.

Like all other short-lived mammals, they have acquired characteristics that allow them to make the maximum possible investment in fertility at a young age.

The advent of trawlers — large ships that fish by dragging, or trawling, a funnel-shaped net behind them — brought about a drastic change in the circumstances in which cod live. Trawlers have become ever more powerful, the nets have become bigger, and their meshes smaller, all in the pursuit of landing ever greater catches. This new kind of fishing means that cod have much less chance of surviving in the sea if they are older, and therefore bigger. Only small, young fish can slip through the trawlers' nets. This has led to a significant drop in the average age of cod living in the sea. Before the advent of trawler fishing, it was mainly the large cod that spawned, as they did so late in life. Now it is the small, young fish that maintain the species' population. Fishery biologists have discovered that cod are reaching sexual maturity at a much younger age and smaller body size than before. The cod has evolved through this (non-natural) selection towards earlier development, and has thus adapted as a species to its changing environment.

Humans and elephants are far less exposed in their natural environments to dangers that could threaten their existence than cod are, and so their lives can last much longer. They can afford to invest in the maintenance and repair of their own bodies. Indeed, such investments are necessary to enable them to produce sufficient offspring, despite requiring many years to develop to sexual maturity, and long gestation periods for their unborn young. If we compare the lifetimes of mice, humans, elephants, and all other mammals, we see that their average life expectancy is inversely proportional to the number of offspring they produce. That is: the more offspring, the shorter the life. The lifespans of different species can vary from short and sharp to long and languid. But natural selection has given each species the length of life that fits the environment it lives in. Nowhere in the animal world do we find the combination of a perfectly maintained body and many offspring. This is in accordance with Tom Kirkwood's disposable soma theory, and the distribution of scarce resources.

Some species' lifecycle seems extremely strange, but can be better understood if we bear the disposable soma theory in mind. Salmon live most of their lives at sea. When they have built up enough fat reserves for migration and spawning, they enter rivers and begin to swim upstream. Once in the freshwater environment, they undergo a metamorphosis. Their colour changes, and the males develop a hook — called a kype — on their lower jaw. The spawning period lasts for about two weeks, and is so intense that most salmon do not survive it. In a very short space of time, they put all means available into their fertility and reproduction, at the cost of their own survival. As the season ends, the spawning grounds are strewn with dead salmon. Fewer than 5 per cent of adult salmon will eventually make it back to the ocean, where they return to their salt-water form and continue to grow. Their progeny will follow in their wake. And one day, they, too, will swim upstream and repeat the cycle.

But that is not even the most extreme example. After successful mating, the female praying mantis devours her partner. In this way, the male provides food for the female, to the benefit of both their offspring.

THE COST OF SEX

In the pursuit of experimental evidence for the disposable soma theory, much research has been done on fruit flies. These insects have been used for many years in a huge range of biological investigations, and have provided scientists with a wealth of knowledge. They live only a few weeks, which is an important prerequisite for studies into animals' lifespans. Another, not unimportant, advantage is that they are cheap to raise in large numbers in the lab. However, scientists face a fundamental problem when carrying out their ageing experiments: when they identify a fruit fly that has reached a great age, it will, of course, be biologically old, less fertile, and on the verge of death. But it is precisely these animals the researchers want to use for their (crossbreeding) experiments.

During his doctoral research, Bas Zwaan, an inspired Dutch professor of genetics, came up with an elegant solution to this problem. He made use of the phenomenon that fruit flies live longer when the temperature is low, and shorter when it is high. He divided each new generation of flies randomly into two groups. One group was exposed to a high-temperature environment, allowing the scientist to establish quickly which of the otherwise indistinguishable individuals were naturally long-lived. Once he had found out in this way which families were long-lived and which were short-lived, he took their siblings, which had been kept in cooler conditions and so were still alive, to use in his experiments. And his results were remarkable. It turned out that repeatedly selecting the longest-lived flies for reproduction led to a significant increase in average lifespan within just a few generations. Bas Zwaan's selection programme for longevity in fruit flies was comparable to, and just as effective as, the ancient selective breeding activities of dairy farmers in Friesland, leading to the famous high-milk-producing qualities of today's Friesian cows.

At the same time, Bas Zwaan also studied the flies' capacity to produce eggs. Remember, he had selected neither for nor against this characteristic. He found that the long-lived flies produced fewer eggs during their lifetimes. And the opposite also turned out to be the case: fruit flies that produced a lot of eggs lived significantly shorter lives. These biological experimental results are in accordance with the disposable soma theory: investments in living longer come at the expense of investments in the number of progeny produced.

The underlying mechanism in the issue of whether body or brood is prioritised has also been described as 'the cost of sex'. The British biologist Linda Partridge decided to try to get to the bottom of this mechanism. One of her experiments involved studying survival rates among male fruit flies that she had divided into two groups: one group that had sex, and another that did not. She found that the mortality rate among males that were presented with females to mate with every day was five times higher than that among males that were not allowed to mate. This increased risk of dying disappeared again as soon as the males were denied contact with females. Under the microscope, it could be seen that the sexually active males had suffered considerable damage, including to their wings, during the act of mating. Since fruit flies have only a very limited capacity for repair, sex usually leads to lasting damage, a loss of functionality, and thus increased mortality. Such complications related to mating are known as the 'direct cost of sex'.

There are also 'indirect' costs of sex. They are not linked to the act of mating itself, but to the ability to be sexually active and to produce offspring. In other words, this is the price that has to be paid for the ability to make several versions of the same species. Most species have two sexes — male and female — each of which has just one kind of gamete, or reproductive cell. But hermaphroditism — the state of having both genders — also exists in the animal world. Solitary hermaphrodites can reproduce independently, without the need to mate with other individuals from the same species.

We get a first indication of the indirect cost of sex when we look at sexless species such as the hydra. It reproduces by means of budding from one of its 'totipotent' stem cells, which are distributed throughout its whole body. No sex of any kind is involved in this method of reproduction. These are the same stem cells that can be used to repair all types of tissue damage. In this way, the little polyp can reproduce by cloning itself, while managing to avoid the ageing process by means of its excellent regenerative abilities.

Not all that long ago, it was discovered in the lab that some hydras undergo a sexual transformation. This happens when conditions become less favourable. Hydras see such a change as a threat to their survival. The asexual animals react by becoming hermaphrodite hydras. The evolutionary logic of this is that sexual reproduction will give a hermaphroditic hydra more chance than asexual cloning of producing successful offspring in a changing environment. But there is a high price to be paid for this sexual transformation. Unlike their asexual counterparts, hermaphrodite hydras' statistical chance of dying does increase as they get older. Hermaphrodite hydras do age!

Stem-cell researchers have found out why such hydras are susceptible to ageing. In all mammals, the reproductive cells originate in what biologists call the germ line — in the ovaries or testicles — which is made up of 'unipotent' stem cells. These are stem cells that only have the ability to develop into reproductive cells, or gametes. So, the stem cells in the germ line are fundamentally different from their 'totipotent' counterparts that can develop into any kind of bodily tissue and which can be used by a hydra, for example, to clone itself. When a hydra becomes a hermaphrodite, it still possesses stem cells, as in the germ line, but they no longer have the 'totipotent' ability that those of an asexual hydra have. The conclusion is obvious: such a sexual transformation comes at the expense of a hydra's regenerative powers. Or, to put it another way: a hermaphrodite hydra's body becomes 'disposable'.

ARISTOCRATIC FRUIT FLIES

In 1998, I had the opportunity to work with Tom Kirkwood for a year in Manchester. His disposable soma theory inspired me to think about how and why we humans age. Once I had completely understood the evolutionary concepts involved in the theory, I was faced with the inevitable question: Does the disposable soma theory apply to humans, too? Until then, the only evidence gathered on the theory was from fruit flies.

In Leiden, studying under Professor Jan Vandenbroucke, I had been trained in the observational method for the scientific study of humans — called the epidemiological method. Most research biologist use experiments: working in the lab, they deliberately introduce a change in the genetic material of their test animals, or in their environment, and then study the effects of that change. However, for various reasons, most such laboratory experiments cannot be carried out on humans. Jan Vandenbroucke taught me that spontaneous events in human populations can resemble the kind of deliberate interventions made by experimental biologists in their labs, although, of course, those changes were not intended for any particular purpose. By closely studying these events — pseudo-experiments — in human populations, a lot can be learned about the causes underlying illness and health. For example, patients suffering from age-related blindness often turn out to be carriers of genetic variations in the body's immune defences against infections. This knowledge has helped to define the role of immune defences in causing damage to the retina.

I suggested to Tom Kirkwood that we should use the epidemiological method to test his disposable soma theory. His initial reaction was negative. What spontaneous events among humans could be compared to Bas Zwaan's crossbreeding experiments with flies? Unlike in an experiment in the lab, where flies that have been specially selected for longevity are forced to have sex, human beings usually have sex voluntarily and fairly randomly. However, the principle that sexual reproduction jumbles up the genetic material of a male and a female is the same for humans as it is for flies. For that reason, the children produced by a couple can be seen as the results of a genetic experiment carried out by the mother and father

— even if that was not, of course, the motivation of the parents for having those children.

Kirkwood eventually came round to my way of thinking. Using the guiding principle of Bas Zwaan's experiments, Tom Kirkwood and I began searching for sets of parents in which he, she, or both had reached advanced old age, in order to investigate whether the children of that union lived longer than average, and had fewer children than average themselves. As always in scientific research, there was an element of luck involved. In that year, 1998, advertisements appeared in the British newspapers for a genealogical CD containing every aristocratic title, and a list of those who held them. From time immemorial, the British nobility has kept precise family history records — who married whom, how many children they had, and how long they lived. This CD made those archives available, even to 'commoners', with the click of a mouse. Anyone could check whether they were descended from a noble family.

That CD presented us researchers with a golden opportunity. It offered a chance to study the descendants of thousands of married couples as if in a natural experiment. The fact that they were all aristocrats also solved another tricky problem. Differences between socio-economic classes can seriously disrupt observational studies. Members of the upper class have a lower mortality rate, live longer, and have fewer children, on average. A superficial examination of the data can lead to a false conclusion: that class can explain why some people live longer and have fewer children than others. However, since these British aristocratic families have always belonged to the upper class through the centuries, and people from lower socio-economic classes were not included in the archives, this problem did not occur.

Once Kirkwood and I had sorted the thousands of aristocratic lives into groups, we noticed that married noblewomen who died young had had fewer offspring than those who lived longer. That's logical, of course, because a longer lifetime means more opportunity to bring children into the world. But, a long life after the menopause clearly does not have that advantage, since a woman's natural fertility is then at an end. Remarkably, it appeared that noblewomen who reached the age of 80 and above were more often childless or had given birth to just one living child. It makes no sense to suppose that these women had deliberately decided early in life to remain childless or to have few children so that they could live to be very old. On the contrary, succession was of paramount importance in British aristocratic families, and so we can reasonably assume that these women's childlessness was not by choice. In line with the disposable soma theory, and in accordance with the fruit-fly experiments of Bas Zwaan, we were able to conclude that a long human life and a large number of children usually do not go together.

After Kirkwood and I published a scientific paper describing our findings, the British press — always on the lookout for a juicy story — ran the headline 'British Aristocracy Mate Like Fruit Flies'.

One remarkable realisation that came out of our study of the British aristocracy was that in modern times there no longer appeared to be a relation between age at death and number of children — I return to this point in Chapter 5 — as if there were no longer any 'costs of sex' in our modern society. In part, that is, in fact, the case. Death in childbirth, let alone death as a direct consequence of having sex, are now extremely rare. Most sexually transmitted diseases can now be cured. However, the question remains whether having more or less sex makes you live longer or die sooner. It is often claimed that sex is good for your health. However, this is based on the fact that healthy people remain sexually active into old age, while those who report less sexual activity tend to be frail and sick. But is it not probable that people who are sick and infirm have less sex for that very reason, rather than the other way round? Sex in old age has an emotional value, and is important in relationships. It is unlikely that it makes you either live longer or die sooner.

Sexual reproduction forms an integral part of the lives of humans and other (mammalian) animals. This evolutionary development has led us to invest in our offspring at the expense of ourselves. This provides a logical explanation for the reason people age. In the chapters that follow, I further explore the way in which sex is partially responsible for causing us to age.

4

ASSESSORS OF THE FINITE

Rising life expectancy is a headache for managers of life-insurance companies and pension funds. Fortunately, however, our life expectancy can be captured by mathematical formulas. There is an arithmetical model for predicting the finiteness of life. Our risk of dying doubles every eight years, and increases exponentially. Some scientists infer from this that the rate of ageing among humans remains constant, irrespective of the circumstances they live in, and is genetically determined. But apart from our genes, we also have to deal with environmental factors and pure chance.

There has always been a close connection between money and the length of human lives. As early as the seventeenth century, the States of Holland, as the Dutch government was called at the time, sold annuities to help finance social schemes. Citizens paid a purchase price to the government, and in return received a regular income for the rest of their lives. The annuity was usually between 6 and 7 per cent of the sum invested, and this financial product was above all used as a life-insurance policy for widows. Around 1670, the government became concerned about the large financial burden of the payments to be made. Representatives of the States went to The Hague to ask the highest official — known as the Grand Pensionary — Johan de Witt, to work out a more sustainable system. In Amsterdam, the city council appealed to the mayor, Johannes Hudde, with the same request.

Both De Witt and Hudde were not only political leaders, but also mathematicians. Due to his position, burgomaster Hudde had access to the contracts of sale for all annuities sold to the people of Amsterdam by the city. He drew up a 'Table of Mortification' to calculate the number of years lived by insured persons, ordered by age. On the basis of these calculations, Hudde concluded that the annuity payments were too high. Nonetheless, the payments were not changed; there was no support for such a move. There were countless political and emotional reasons to ignore the results of Hudde's calculations. And there still are.

INSURANCE PREMIUM LEVELS

Estimating life expectancies accurately is crucial for determining the associated risks, incomes, premiums, and payments from pension funds and life-insurance policies. In 1693, the astronomer Edmond Halley — of comet fame — drew up a proper 'life table'. For the first time, this table provided a way to infer the average human lifespan. Today, those who make their living doing just that are called actuaries, and they often have a mathematical background. Actuarial science has changed little in essence over the past centuries: actuaries still spend their time drawing up life-expectancy tables. However, unlike their predecessors, modern actuaries also try to predict average life expectancies in the future. Pension funds and insurers need this information to secure their own financial futures. Increasing life expectancy is a fine thing in general, but it is a headache for pension-fund managers and life-insurance companies. How should they break the 'bad' side of the 'good' news to their customers? Is it time to reduce payouts from pensions and policies? Should the premiums be raised, or should pension-plan holders be asked to pay into the system for longer? Or both?

Despite their care and precision, actuaries have a weakness: they are not very good at predicting the future. At times in the past, they have overestimated the development of life expectancy, but usually they have erred on the side of underestimation. Time and again, life expectancy has turned out to have risen more than was ever thought possible. This is one of the reasons that the premiums paid turn out to be too little, and payouts are correspondingly too high.

Almost all current models, scenarios, and projections of life expectancy are based on the ideas of Benjamin Gompertz, a mathematician who came up with a simple arithmetical model in 1825. What he did, in essence, was to divorce the mortality rate due to ageing from the non-age-related probability of dying. We have encountered the latter risk before in the case of the hydra that gets gobbled up by a stickleback; in human terms, it might be the risk of dying in a train accident, for example.

The Gompertz model is very powerful. To this day, it remains a touchstone for scientists trying to express the ageing process in numbers. Imagine that an individual's chance of mortality does *not* increase with the passing years, as is indeed the case for hydras, so that the chance will not have changed in ten or even twenty-five years. Does that make hydras immortal? No, because their chance of dying is always greater than zero. However, since the risk is constant over time, and does not increase with age, there will be little change, in the biological sense, to a given hydra, and the little polyp will not age. Fate, which makes something go terribly wrong every now and then, is a fundamentally different thing to the ageing process, which gnaws away at our bodies, making us frail, sick, and, eventually, dead.

Gompertz also showed how much an individual's risk of mortality increased with age. This increase is exponential — that is, it doubles every eight years. That is as true for 23-year-olds as it is for 57- or 83-year-olds. Note, however, that an 83-year-old's chance of dying at any given moment is higher than that of a 57-year-old's or a 23-year-old's. This constant rate of doubling also appears to be true not only for people in the UK, who formed the initial basis for Gompertz's model, but also for people in the Netherlands or Africa, for example. And the model is not only applicable to Gompertz's time, but also to the present, and even to periods of famine and war. This does not mean that mortality rates are the same in all countries and under all circumstances. Indeed, mortality rates vary hugely. Those variations are determined by the time, the environment, and the conditions that people live in. But over and over again, it has been found that the rate at which the risk of mortality doubles always remains the same.

A number of scientists conclude from this that the rate of ageing among humans remains constant under any circumstances. They consider this to be an idiosyncratic feature of the ageing process. They see it as the result of a long process of natural selection, and as a genetically determined process that is untouched by outside influences.

The Gompertz model also applies to the mortality rates of mice, elephants, and other animals. Mortality among all mammals, without exception, increases exponentially with age. The time it takes for the risk of dying to double is the shortest for mice; they age quickly. The doubling time is the longest for elephants; they age slowly. Humans are somewhere in between. This means the rate at which the ageing process progresses can be described with a single number. It varies from species to species, but for individuals of one species, it is always the same. This explains why scientists get so excited about the value of this constant rate of doubling. Although the mortality rates of mice, elephants, and humans are characterised by a constant rate of doubling, that number tells us nothing about the biological explanation for this phenomenon. In fact, the causes of ageing vary greatly between different species.

It is interesting to note that the chance of a car or a washing machine breaking down or having to be replaced also increases exponentially with age. In other words, the chance of failure for complex machines also doubles at a constant rate over time. This means we need a whole new way of thinking if we are to understand why this rate of doubling should always be constant. All things considered, it appears that the rate of ageing is governed by some sort of law: the chance that any complex system will fail always increases exponentially over time. In Chapter 8, I use the biological mechanism underlying ageing in humans as the basis for demonstrating that accelerations in the accumulation of damage can explain this constant rate of doubling. And that is true for organisms and machines alike.

THE IMPOTENCE OF PREDICTION

It is a fact that your father and your mother partly determine your chances of getting sick or staying well, by means of the genetic material they pass on to you. They give you resilience to damage and infection, and the capacity for repair, but they also burden you with imperfections, faults, and defects. This might sound strange, but we saw in Chapter 3 that natural selection does not lead to all characteristics evolving to perfection, since that is not necessary for the perpetuation of the species. Since we receive a mixture of good and not-so-good qualities from our parents, no one is perfect, and almost everyone is about average. If we want to enjoy our average body and mind for a long time, we need to look after them. Some people manage to turn their body into a wreck within a just a few years, with a wild life of sex, drugs, and rock'n'roll. That ages their body rapidly, as it accumulates defects early in life, thus greatly increasing the risk of sickness and death.

We already know a great deal about avoiding that sort of damage to the body. Don't smoke, and don't eat or drink to excess, is the most common advice. And, of course, take sufficient exercise. That is how to keep your chance of mortality low.

Smoking is an external cause of damage to the body, and so would seem to be easy to avoid. By the same token, it should be equally easy to eat less and not get fat. But real-life experience teaches us that this is not the case. Many people really struggle to keep their weight down, or to drink in moderation. This is not very surprising, since all these habits have a strong genetic component. The age at which a person has his or her first alcoholic drink is mainly determined by the environment — in this case, the social situation. But there is a genetic background to whether that person stops after one or two glasses, or keeps on drinking. In the same way, a tobacco addiction is easy to kick for some, but cannot (easily) be overcome by others. This is also, to some extent, determined by our genes.

Contrary to popular belief, risk factors such as high blood pressure and high blood cholesterol levels are also largely 'inborn'. In fact, cholesterol levels are mostly determined by genetic factors. However, it is wrong to think that nothing can be done in the face of such a genetic predisposition. It is true that our genetic code is fixed in our DNA, but high blood pressure and high blood cholesterol, or even addiction to cigarettes or food, can be treated or prevented. The options include diets, medication, will power and perseverance, and personal coaches. Just because you have a genetic predisposition does not mean you must resign yourself to the fact that the course of events is inevitable.

And sometimes it is impossible to understand why something happens. You might have good genes and never have smoked, and, suddenly, you pick up a deadly virus, catch a multi-resistant bacterial infection from your food, or develop an irregular heartbeat. The list of the things that can go wrong seems endless, so that it is impossible to predict when someone will get sick or die, even if they have taken extremely good care of themselves. Doctors are no exception. They can diagnose illness, but they cannot accurately predict how long a person will live.

It is often said that a third of the variation in our life expectancy can be attributed to our genes. It is claimed that another third is determined by our environment, the conditions we live in: rich or poor, smoker or non-smoker, sporting ace or couch potato, and so on. The final third is put down to chance. But we really should not seek to separate these three causes from one another, as if one particular illness is genetic, another is your own fault, and yet another is just random bad luck. Sickness and death are the result of the failure of a complex system. In every case, both genetic and environmental factors contribute to longevity and the age we die at.

5

SURVIVING IN HARSH CONDITIONS

Species arise and disappear in a continuous process without a preconceived plan. Survival must have been difficult for the first humans living in harsh conditions. We can study the processes of natural selection and ageing in their original form in Ghana, where people still live under the same severe conditions as our ancestors. Examining those processes provides us with important insights into resistance to infectious disease, the link between fertility and the immune system, the survival of women beyond menopause, and old age in men.

Several years ago, I was invited to give a talk in New Zealand on evolution and ageing. After my presentation was over, I was walking on the beach when I stumbled upon a monument marking the spot where humans first set foot on New Zealand soil. Historians believe they have found evidence that those humans made the crossing to New Zealand from the islands of Polynesia in canoes. Those adventurers must have been the first to survive the long journey of thousands of kilometres. They were called Maoris, and are considered the original inhabitants of New Zealand. The inscription on the plaque has stuck in my mind: 'There are no signs of human habitation in New Zealand prior to the year AD 900.' That makes New Zealand the last landmass of any size to have been populated by humans. It is a far-flung corner of the world; it had taken me 24 hours to fly there from Amsterdam. But a single day is nothing compared to the millions of years it took us as a species to complete the gruelling journey from the middle of Africa, via the Polynesian islands, to New Zealand.

The further you look back into your family, the bigger it gets. This means that two (randomly chosen) people must always have some ancestors in common. One evening, I received a phone call from a Westendorp whom I did not know. He knew that I was married, and he was able to name my two daughters. The caller turned out to have carried out extensive genealogical research on our family history. He was hoping I could provide some information he needed to fill a gap in our family tree. Unfortunately, I was of no use to him; I simply did not know enough. It is not unusual to have an inkling about the identity of your grandparents' parents, but what about their brothers and sisters? I daresay most people know nothing about their family four or more generations back. The purpose of a family tree is to fill that information gap — to trace your roots and your relations, sometimes for several generations, sometimes much further into the past.

Somewhere along the line, I must be related to the British nobility, and even to the Maoris of New Zealand. Finding that relationship may, in the most extreme case, mean going back 250,000 generations. But there is no doubt that I can trace a single unbroken line of copies of my DNA all the way back to the middle of Africa.

AN EXTRAORDINARY FIND IN CHAD

Scientists who study the family tree of the human species are called 'paleoanthropologists'. In order to reconstruct a family tree going back millions of years, they have to look at fragments of bone, teeth, and skulls, which can often be difficult to interpret. This means there are many open questions in this science: which species lived alongside each other; where and when did one species turn into another? It is no surprise, then, that a new find can turn ideas about lineages and relationships on their head. In 2002, a skull was discovered in northern Chad, which many believe belonged to the oldest human ancestor found so far. Proving a claim like that is difficult; after all, what constitutes 'human'? If you were to meet this oldest of human ancestors walking down the street, you would probably be inclined to think it had escaped from the zoo. This new species was given the name *Sahelanthropus tchadensis*, meaning 'Sahel man from Chad'. It is purported to be the ancestor of various later 'hominid', or human-like, species, of which the human species is the only survivor.

S. tchadensis is thought to have been a tree-dweller that was common 6 to 7 million years ago in West Africa, and lived alongside the ancestors of modern apes. Some paleoanthropologists point out that the geographical location of this find is very far from the areas where most other human fossils are uncovered, in eastern and southern Africa. For this reason, they call the mainstream interpretation of this new find into question.

The systematic classification of plants and animals as developed by biologists places the great apes — which are large and tailless — as our closest relatives. The foundations of this classification system were laid down by the Swedish physician and botanist Carl Linnaeus, who gained his doctoral title from Hardewijk University in the Netherlands, and lived and worked in Leiden from 1735 to 1738. Linnaeus, who still believed in the divine creation of natural species, grouped plants and animals according to their outward characteristics. His method was very similar to that of modern-day paleoanthropologists, who have to make conclusions on the basis of bones, teeth, and skulls. Now, most scientists believe species are the result of evolution, and prefer to order them according to their descent — that is, their genetic relatedness. The classic Linnaean system often agrees with a classification based on genetic relationships, but not always, by any means.

Not everything that looks the same has the same origin. For example, we now know that, despite their remarkable similarity, swifts and swallows are not closely related to each other. Swifts are related to hummingbirds, and are not close to the passerines — *or songbirds* — which is the group that contains swallows. Another example is that of the auks in the northern hemisphere, which are as alike as two peas in a pod to the penguins of the southern hemisphere. This similarity should come as no surprise, since the physical circumstances are comparable in both polar regions, and necessitate comparable evolutionary adaptations for survival. This phenomenon is known as 'convergent evolution'. But auks are not related in the slightest to penguins, and have more in common with gulls, terns, and skuas.

Our rapidly increasing ability to determine variations in DNA at the tiniest level, and the ever-increasing capacity of our computers to carry out complex analyses, are a great help in understanding patterns of descent. Thus, we now know that humans and chimpanzees share 98 per cent of their DNA. And we know for certain that humans are not descended from chimpanzees. Genetic similarities *and* differences allow us to conclude that humans and chimpanzees share a common ancestor. It must be the case that the earliest human-like creatures and the earliest ape-like creatures differed only very slightly. However, those differences increased with time. In the course of millions of years of evolution, we humans have learned to think conceptually, to talk, and to build machines.

Note that we are not the only species to have evolved in that time from those original tree-dwellers. Bones and skulls show that there have been more than twenty different hominid species in the course of human evolution, including such well-known characters as Java Man and the Neanderthals. However, all those other species have died out. Apparently, the others were not able to overcome the environmental challenges they faced. *Homo sapiens*, the species we belong to, is — as yet — the only exception. The appearance and disappearance of species is a constant and unstoppable process, and it is most probable that our own species will one day become extinct.

Surviving in the adverse conditions in Africa must have been a constant 'struggle for life' for those early hominids. Just like us, our early ancestors had no idea what the future would bring, or how they should prepare for it. There is no predetermined plan: evolution does not think rationally. Lives simply run their course. Only those newborn babies that manage to survive, reach adulthood, and have children can pass their hereditary characteristics on to the next generation via their genes. Since the chance of survival is generally limited, individuals can never invest enough in their offspring. Even today, death and extinction lie just around the corner — for you, your family, or the entire species.

Because it does not follow a predetermined plan, the erratic course of human evolution can only be understood in hindsight. Family trees are incomprehensible if you follow them upwards, from root to branch. Over and over, you end up following extinct branches, forcing you to go back and start again. Then, suddenly, the lineage you are following leads all the way through human evolution to the present day. It is as if it were suddenly obvious which route takes you to the centre of the maze, where a prize awaits you. Trying to work out the correct route beforehand is absolutely pointless.

The evolutionary development of the human species is a tangled web. Each generation produced offspring, but only some of those individuals will have descendants alive today. Many efforts led to nothing because the newborns were not well adapted to their environment, happened to die young, or failed to produce offspring of their own. Constantly producing more and more identical offspring will not help.

Sex, however, does help increase the survival chances of a lineage. Sex is a fantastic mechanism for creating new variants of the same species. This occurs in a two-step process. First, father and mother split the genetic material to be passed on — the number of chromosomes — in two, through a process called 'meiotic cell division'. This occurs in the reproductive cells. The resulting egg or sperm cells each contain half the total number of the required chromosomes. When the sex cells merge during fertilisation, the material is recombined. This results in a reshuffled, complete set of chromosomes, and life can continue to develop.

Sex leads to the creation of countless new variants, and these can eventually evolve into new species. Some offspring that are the result of sexual reproduction are likely to be better equipped to survive in a new environment than those that result from asexual reproduction and are therefore identical to their parents. Looking within our own species, we may think of Java Man and the Neanderthals as examples of such variants; initially, they were successful as species, but they ultimately disappeared nevertheless. Other examples are aristocratic lineages in Britain, to which the current royal family, the House of Windsor, is a lucky exception; the dynasty has successfully managed to survive so far. Darwin was right when he remarked that it is only the survivors who compete in the struggle called 'fitness'.

However, there are few people whose motivation for

having sex is a conscious desire to perpetuate the human species. We have an instinctive urge to have sex, and it is usually done in the pursuit of pleasure. Both the urge and the pleasure involved in satisfying it are genetically determined, fixed in the code of our DNA. Sex is an evolutionary adaptation that has enabled us to survive to this day as a species that reproduces sexually, albeit to the detriment of our own bodies. It is no surprise that the British biologist Richard Dawkins gave his seminal 1976 book the title *The Selfish Gene*. Our entire genetic programme is geared to favour DNA, not humans themselves. We are nothing but temporary containers for a molecule called life, and the only thing that is thrown away is the packaging — which is us.

The evolutionary rat race — which species will survive and which will die out? — is a fascinating phenomenon, and a good understanding of it is necessary if we are to get to the bottom of the question of why we age. Investments in development, sex, and reproduction have been strictly determined by natural selection, and are encoded in our DNA. As time passes, the costs of that programme accumulate, and that explains why our bodies deteriorate. Within an evolutionary framework, the development of children and the ageing of adults are two sides of the same coin. This is an essential insight for medical professionals.

If you ask a group of aspiring doctors what area of medicine they would like to specialise in, around a quarter of them will say paediatrics. That is a lot, since there are not that many sick children in developed countries these days. This desire to become a paediatrician is an expression of a genetically determined trait that raises the human species' level of fitness. We are genetically programmed to love children. Only very few student doctors say they want to work with old people after they qualify, although doctors' waiting rooms are thronged with droves of old patients. There is simply no evolutionary pressure to select for a love of old people.

For someone like me, who is responsible for training doctors in geriatric medicine, such student aspirations can be pretty frustrating, until you realise how our brains are wired. If people are likely to neglect their bodies in old age, as the disposable soma theory predicts, why would student doctors choose a medical career in gerontology? Is it logical to conclude that they would want to work with older people, just because everyone reaches old age these days? Of course not.

Knowing this evolutionary background, it is easier to understand the choices made by today's medical students, and this has enabled me to develop a more realistic training programme. In 2000, I decided to change tack completely; I began teaching them about sex, reproduction, and evolution. I showed my students that ageing is a logical consequence of all those things. That set the ball rolling. As broad-based professionals, many of my students later forged careers in geriatric medicine.

THE GOLD COAST OF AFRICA

In 2002, a group of researchers from the Department of Gerontology and Geriatrics at Leiden University decided to study the ageing process among people in Ghana. Our motivation was scientific curiosity, but also, and more importantly, our aim was to gain a better understanding as medical doctors of the way our bodies and minds are structured, and how they eventually deteriorate. This meant we had to study the human body and mind in the unforgiving environment they originally evolved in — an environment of excruciating temperatures, food shortages, and infectious diseases — where medical intervention is not the order of the day. Those times are long gone in the West, and so we cannot study natural selection and the ageing process in their original form at home. However, there are still plenty of places in north-eastern Ghana where the conditions still pose a constant threat to the people who live there. So that is where we chose to carry out our ten-year study of young and old people.

The colonisation of Ghana, formerly the Gold Coast, began with the arrival of the Portuguese in the fifteenth century. They were followed by the Dutch in the seventeenth century, who were in turn followed by British overlords. Today, Ghana is a successful west African democracy. However, our research area in the north-east is still very 'primal'. The area has remained practically untouched to this day, for the simple reason that it has virtually nothing to offer. Food is scarce, and the average income is estimated to be one dollar per person per day. There is no money for artificial fertilisers, and the rugged climate means that crop failures are common. People there work the land with traditional tools such as hoes. Extremely rarely, you might see an ox. Motorised agricultural machinery, such as tractors, does not exist there.

The Bimoba tribe, whose society is patriarchal, have lived there for centuries in a very primal way, close to nature. The head of a household lives with one or more women, including his mother if she is still alive, in a row of thatched huts. The huts are encircled by a wall, forming a compound. The bodies and minds of these people are adapted to this environment, which is characterised by scarcity and danger, and that is precisely why we chose to go there as researchers from the West.

We began by 'mapping' the area in a medical-anthropology sense. Who are its inhabitants? How many children do mothers tend to have? How many do fathers have? And how do these people live? The Bimoba were a blank slate as far as the scientific literature was concerned. As expected, the pattern of mortality on the ground was dominated by starvation and infection, with prosperous years alternating with periods of scarcity. Practically everything in the region is reddish-brown and dusty; precipitation and growth appear only during two short rainy seasons per year.

Arriving in the region as a Western researcher, it appeared at first glance that poverty pervaded everything. Later, we began to notice subtle signs of rank and position: a pig perhaps, and the occasional rectangular hut with a corrugated iron roof and an electricity connection. Such attainments indicated higher socio-economic status. And, as they do everywhere and all the time, these differences in wealth translate directly into a higher or lower risk of mortality. Accordingly, we found that the people with the lowest socio-economic status were twice as likely to die at any given time as the highest-status individuals. A higher income offers the possibility of more food, of vaccinating your children, of digging a well for clean drinking water. Money raises individuals' life expectancy, because wealth offers more possibilities for survival in a hostile environment.

From this it is clear that the economic principle of the distribution of limited resources is fully applicable to this community in north-eastern Ghana — not only in the real economy, but also in human life-trajectories. Here, too, the disposable soma theory stands up: these people must choose between investing in reproduction and investing in maintaining their own bodies. Having children and bringing them up requires great (financial) investment on the part of the parents. But, at the same time, children are an insurance policy for a parent's old age. However, when conditions are bad, one set of parents in every five in Ghana will lose all their children before they reach adulthood.

This is probably the reason that having a large family is a source of prestige and status. Men aspire to the taking of as many wives as possible. Each wife requires investment in the form of a dowry, which only a few men can afford. Generally, all women have a large number of children, and childlessness is a social taboo. The number of children a man has within the same family can reach thirty or even forty. There are also men in this polygamous society who cannot afford a wife. They are excluded from the evolutionary game; the number of offspring they father, notwithstanding extramarital sex, is zero. And so everyone is part of the subtle game called 'fitness', in which human behaviour, property, and reproduction all play a role.

Over many generations, the Bimoba have developed a culture that enables them to survive in this unfavourable environment. Their entire social order is geared towards prioritising having children, and so maximising the tribe's chances of survival. The limited resources on which the local population depend for a living have left a lasting mark on their social structure and in their genetic material. In this polygamous society, it is the dominant and wealthy compound landlord, the male head of the household, who fathers the most children. Parallels with the alpha males of primate societies are obvious. The dominant and submissive behaviour of apes, the strict hierarchy within the group, grooming behaviour — everything that evolutionary biologists study — has a strongly genetic background. Just like apes, humans are full of emotions that have an evolutionary basis. Our behaviour is coloured by the environment we live in, but the 'hardware', the character traits that underlie that behaviour, is also determined to a great extent by specific genetic characteristics. This is the human behaviour that has been selected for in the polygamous society of the Bimoba.

In various places around the world, scientists research the genetic basis for human behaviour. They often focus their studies on the differences between identical and fraternal twins. Identical twins share all their genes, while fraternal twins share half, just like any two siblings. By studying the differences between these two types of twins, it is possible to determine which human traits have a genetic basis and which do not. Dominant behaviour, tenacity, intelligence, but also sociability, affectionateness, and optimism are determined to a great extent by our DNA. All these skills are necessary to gain and maintain a place in society. This does not mean that everyone possesses every trait in the same measure. Although the average height in our society may be 1.8 metres, some people measure 1.6 metres; others; 2 metres. Such variations — whether greater or smaller — are all part of the normal range of variability.

Natural selection has adapted the way we function

socially to the environment we live in, and, conversely, we attempt to mould the environment to suit us by means of our behaviour. We are not simply slaves to our genes. People in Africa can choose not to get married, or to remain childless. Such decisions are influenced by their upbringing, by the people they associate with, and by the people they turn to for advice. And, of course, their behaviour also depends on their circumstances and the (in)tangible resources available to them. Does a man have enough money for a dowry, so that he can marry a woman?

The life trajectories of those of us who live in modern developed societies can be explained by our parentage — what genetic basis did I inherit from them? — and by the circumstances we live in. It is nonsense to try to explain the biological events in our lives, such as illnesses, and our behaviour, which is based on our emotions, on the basis of our genetic makeup or our environment alone. Both are important. Even someone who has been an optimist his whole life can be so upset by the death of his beloved wife that he slides into depression. And the opposite is also true. Someone with a rather bristly personality who has never been able to form an attachment to a partner throughout her life can find herself falling head over heels in love, and getting married in old age.

RESISTANCE TO INFECTIOUS DISEASE

The women in our research area in Ghana gave birth on average to between six and seven children. When circumstances are harsh, it is necessary to produce that number of offspring to guarantee a constant population in the long term. If women have even more children, the survival rate of their offspring begins to drop. The first and most obvious explanation for this is that there is less food and fewer (financial) resources available per child. But starvation is not the only life-threatening danger in Ghana. Infections are also an ever-present hazard. If we look back at the evolutionary development of the human species, we see that the risk of contracting an infection increased around ten thousand years ago, when we gave up a hunter-gatherer lifestyle in favour of a settled life in social groups, working the land, and keeping livestock. This change meant we had access to a better and more reliable source of food, but it also brought with it a sharp rise in the risk of infections spreading from animals to humans, or from humans to humans. In response to such infections, humans developed a versatile immune system to keep invaders in check.

Only those offspring with sufficient defences against infection survived the transition from hunting and gathering to agriculture and animal husbandry. The development of agrarian societies progressed in fits and starts. There are countless examples of communities that were wiped out by epidemics. Sometimes, a few individuals managed to survive the catastrophe, and they became the founders of new, thriving communities. This is how modern humans gradually developed their powerful immune defences against infection.

The body's resistance to infectious diseases takes a fundamentally two-pronged approach. The first line of defence is a lookout function. Its job is to identify pathogens as hostile agents. Pathogens do everything they can to remain unnoticed in the human body. They disguise themselves, making it difficult for them to be to identified as foreign bodies, so that they are not targeted by the immune system. This is how certain tropical worms, once they enter the human body, can remain in the bloodstream for thirty years or more. Other pathogens are able to change their disguise repeatedly, so our immune system is taken by surprise again and again, and we get sick every time we encounter them. The flu virus, for example, is slightly different every season, which explains why it can infect so many people each year.

Neutralising pathogens is the second part of our immune defence system against infection. Once the intruder has been identified as alien and hostile, it must be killed and removed. This is achieved by means of an inflammatory response. In the case of flu, inflammation is associated with the familiar symptoms: aching muscles, fever, fatigue. Another example of inflammation is that familiar red, painful, throbbing finger that only gets better when the festering splinter is removed.

It seems logical that the stronger an individual's immune system is, the better, since then her chance of survival is the greatest. So it is quite remarkable that human beings are still susceptible to infections, despite so many generations of selection in favour of resistance. We have managed to acquire other biological traits through natural selection in a very short space of time. For example, when humans began herding livestock, milk became part of our staple diet. By means of natural selection we developed the ability within just a few generations to produce the enzyme lactase, which enables us convert lactose (the sugar that is so plentiful in dairy products) into ordinary, more easily digestible sugar.

It is only when we look at human lives as a whole that it becomes clear why further selection for resistance to infection becomes problematic. If recognition mechanisms for pathogens are never allowed to fail, and inflammatory responses can never be too powerful, undesirable side effects can develop, such as the rejection of an embryo from the womb.

An embryo that needs to stay in its mother's womb long enough to grow into a full-term baby is an equal mixture of its father and its mother. Of course, the mother thinks it is *her* child, but that is not quite the case. Fifty per cent of its genetic material has come from the father, and those genes have an impact on the child's development from the outset. A direct consequence of this is that the mother's immune system has a tendency to identify the embryo in her womb — half mother, half father — as a foreign body, leading it to want to reject the embryo in a 'spontaneous abortion'. To prevent this from happening, proteins are present at the interface between the placenta and the womb that prevent the immune cells of the mother and the child from attacking each other. Nevertheless, such an immune response does sometimes occur, and the result is a pre-term birth. In such cases, the differences between mother and child are too great, and her 'tolerance' too little. The chance of this happening is greater if the mother's immune defences are 'stronger'.

In our study area in north-eastern Ghana, we were able to identify the DNA variation that explains hereditary differences in immune defences between different human individuals. Those who carry certain variations are better equipped to fight off infections and survive them; but, at the same time, pregnant female carriers of that variation have an increased risk of rejecting their embryo. This shows why evolution cannot simply go on forever selecting for ever better defences against infectious disease: the resulting reduction in reproductive fitness leads to those evolutionary variations dying out. Conversely, a 'weak' immune system cannot be selected for limitlessly, simply because that reduces the risk of pre-term birth. If a woman falls prey to a fatal infection before she can ever become pregnant, there can be no question of fitness at all. This results in a constant wavering between the two extremes. This is what we call 'balanced selection'.

This mechanism can offer a better explanation of our previous findings concerning infant mortality in the research area. If the children of mothers who have borne many babies die of infections more often than the average, a lack of sufficient food and money is only part of the reason. The fact that they have a more-than-average susceptibility to infectious disease is a sign that their immune system is less effective at fending off pathogens. Their less 'strong' immune system is inherited from their parents. It is this very genetic makeup that made it possible for their mother to have so many babies.

Our findings were not totally unexpected. This is an example of what scientists call the quality-quantity trade-off — that is, the fact that members of a species will produce either a few offspring of high quality, or many offspring of lower quality. This phenomenon is found throughout the animal and plant worlds, and is in line with the disposable soma theory. In this explanatory model, fertility — *quantity* — is exchanged for sustainability — *quality* — as evolutionary choices are made about which properties should be afforded more investment than others.

This discovery also provides an answer to another question. When Tom Kirkwood and I were studying the British nobility, we were puzzled to find that there appeared to be no correlation between the number of children born to modern-day aristocratic women and their life expectancy. This can be explained in the context of an immunological quality-quantity trade-off. The cost of exchanging fertility for a weak immune system is much lower today. The chance of dying from an infectious disease has been reduced to a minimum. To put it another way, investing in sex and fertility has become less expensive.

THE BENEFIT OF GRANDMOTHERS

Women have a remarkable life trajectory. Girls develop into young women, at which point they become extremely attractive to men. Their fertility reaches its highest when they are in their twenties, then declines rapidly. When they reach the age of 50 or so, their fertility comes to a definitive end when menopause sets in. However, post-menopausal women still have considerable chances of surviving, and many become grandmothers. The remarkable thing about this is that their ovaries are already 'dead', while the rest of their bodies and their minds may well still be perfectly fine. From an evolutionary perspective, the fact that women survive far beyond childbearing age appears to defy logic. According to the disposable soma theory, the survival of women after the menopause should be seen as an investment in longevity at the expense of fertility. The mortality rates for women in old age are actually lower than those for (still fertile) men, which accounts for the fact that most women outlive their male partners. That seems crazy, too. After all, there is no evolutionary pressure to select for longevity, is there? So how can we explain this?

Grandmothers often play a distinctive role in young families with growing children. Grandmas look after their children's children, increasing the (survival) chances of those offspring by doing so. This is the case in modern families, where they provide mainly intangible advantages such as attention and education; but when conditions are severe, the presence or absence of a grandmother may mean the difference between life and death. When a grandmother provides childcare, her daughter is freed up for the production of offspring, and the number of 'copies' of grandmother's genetic material in subsequent generations will increase. This 'grandmother hypothesis', as it is called, explains why women enter the menopause when they are about 50 years old. Since they can no longer produce any children of their own, postmenopausal women are free to fulfil the useful role of caregiver for their grandchildren. In evolutionary terms, postmenopausal survival was selected for because it contributes to the fitness of the species.

A clue to the evolutionary importance of grandmothers in contemporary society can be found in the work of the Dutch sociologists Fleur Thomese and Aart Liefbroer. They studied the involvement of grandparents in the care of young children, and the impact this has on subsequent births in families where both spouses are in paid work. They used data from three generations of men and women in contemporary Dutch families. Their findings showed that maternal grandmothers provided childcare more often than paternal grandmothers did, and that grandmas took care of their grandchildren more often than granddads did. Involvement of grandparents in raising their grandchildren increased the likelihood of more children being born in that family. These results led the researchers to conclude that, from an evolutionary standpoint, childcare by grandparents can indeed be seen as a successful reproductive strategy. Families in which grandparents play a part in the children's upbringing have a greater degree of fitness in our modern age. The outcome of this study is an excellent example of how the processes of natural selection and evolution are not confined to the distant past or faraway Africa, but also apply to us, in the here and now.

The influence of grandmothers on the survival of their grandchildren can no longer be examined in contemporary Dutch society. The favourable conditions we now live in mean that mortality rates among small children are practically zero. This, in turn, means that death in early childhood is no longer a factor in natural selection. So in order to study the impact of grandmothers on infant mortality, researchers have to turn to historical church records. These registers of births, baptisms, marriages, and deaths have been a useful tool for reconstructing the histories of families all over the world. It is a hellish job, due to the major problem posed by the fact that individual people may be recorded in different registers, and those records are not cross-referenced. Some people lived their entire lives in one parish, where all their rites of passage took place. But others moved around during their lives. What about their histories? This type of research requires a long period of data collection and collation in advance, with genealogists and archivists playing a crucial part.

Much of the historical data used in such 'biodemographic' research comes from Scandinavia. In the eighteenth and nineteenth centuries, stable communities arose there, which were connected by their shared religion. It is not difficult to imagine how these communities were structured: small groups of farmsteads clustered around a church, but otherwise cut off from the outside world by topographical features such as rivers, mountain ranges, or ravines. Life must have been hard, with long winters, failed harvests, and grinding poverty. Child mortality was high, and many women died in childbirth, as a late but nonetheless 'direct' result of having sex.

On the basis of the life histories of these people, researchers have been able to show that the presence of a grandmother did play an important role in raising the chances of their grandchildren's survival. It is also not difficult to imagine that, for families living in such conditions, the availability of help when a mother dies in childbirth can be a key factor. Under these circumstances, the grandmother hypothesis is confirmed.

On an evolutionary scale, however, those Scandinavian church communities are a mere drop in the ocean. It is unlikely that a few centuries of natural selection in the developed world can entirely explain the genetic background to the remarkable development in women's lives that is menopause. This female peculiarity must have its origins in a much more distant past, when families were structured very differently. Anthropological descriptions of past societies show that the structure of the overwhelming majority of them was polygamous. New techniques of DNA analysis confirm that, in the past, most men had more than one sexual partner. The position of (grand)mothers in a polygamous relationship is very different from that in a monogamous marriage, and it can have a powerful impact on the results of natural selection.

To really understand women's lives, to get to the bottom of the function of the menopause and to shine a light on the benefit of grandmothers, we would need to carry out a statistical analysis of the birth rates and death rates of the members of polygamous families. However, the historical records necessary for undertaking such work do not exist. Instead, my research team studied the present-day life histories of the Bimoba tribe. Its polygamous structure, high mortality rates, and large number of offspring appear to reflect quite accurately the circumstances under which humans evolved. We reconstructed countless family trees, and determined women's fertility rates and the survival chances of newborn babies.

For mothers in north-eastern Ghana, there is a considerable risk of dying in childbirth. It appeared that the presence or absence of the landlord's mother, the grandmother of his children, had no impact either on the children's survival chances or the number of children that were born. In a polygamous family there is a considerable age difference between wives — a man usually marries several young women over an extended period of time — and when one of them dies, any of the other wives can step in to provide childcare. The wives' mothers played no part in the family structure; they often lived far away, in the women's native villages. The conclusion from our study was undeniable: the presence of a grandmother had no significance for the number of children born in a family, or for the survival rate of those children.

The conundrum was how to reconcile these findings about the life histories of the Bimoba tribe with those based on the Scandinavian church records. The circumstances in which the two sets of grandmothers live(d) can be considered more or less the same, but their social positions were not comparable. Scandinavian grandmothers played a completely different part in the affairs of the family than the mothers of Ghanaian landlords. The role of a grandmother is much greater when a woman in a monogamous marriage dies.

What is beneficial in one environment is not necessarily beneficial in another. Our life trajectories in particular are the result of natural selection within polygamous societies that had to survive in a world of scarcity and ever-present physical dangers and infectious hazards. And that is precisely when the grandmother hypothesis turns out to be an inadequate explanation of the strange postmenopausal lifespan of women.

An alternative explanation has been proposed recently — taking into account the fact that men also exist. Unlike women, with their more or less abrupt menopause, male fertility declines much more gradually. Men often father children well into old age, and that is certainly true of a wealthy Bimoba landlord with the money to take several wives. We calculated that around 20 per cent of Bimoba children were fathered by men over 50. Other researchers have identified comparable percentages in similar circumstances elsewhere. Women cannot become pregnant spontaneously at the age of 50 and above. By contrast, men who father children in old age are still very much competitors in the evolutionary rat race. Since these men have more children on average, they leave an extra-large number of copies of their genetic material behind

— they have an above-average number of offspring, so their fitness is greater. But in order to do this, they must reach a great age in the first place. These old fathers pass on the attribute of longevity to their sons, but also to their daughters. This is an alternative, plausible explanation of the fact that women continue to live for a long time, even when they have stopped investing in being fertile, and have entered menopause.

6

OUR INCREASED LIFE EXPECTANCY

Thanks to a combination of improved hygiene, clean drinking water and better food, control of infections, technical and medical innovations, and less violence and good public governance, the average life expectancy in the developed world has doubled, from 40 to 80 years, in the space of just a century. Today, people die from diseases that appear to be new, but which have in fact always existed. They previously went unnoticed because very few people survived into old age. There is no maximum age encoded in human DNA. Or, to put it another way, every week our lives are extended by the equivalent of a weekend, and there is no end in sight to this development. Ageing is avoidable, but because we will always face a risk of dying by some other means, humans will never be immortal.

The harsh living conditions that still prevail in north-eastern Ghana today are far less removed from us in time than we think. In 1830, cholera swept over the whole of Europe, having spread from India via Russia. The Dutch city of Leiden saw no less than seven outbreaks of the disease between 1832 and 1866, which claimed a total of around five thousand lives. Such events continued to occur in Europe until well into the nineteenth century. This was a period of particular investment in agriculture in the Netherlands. Steam power was being used to reclaim land from the sea, and increasing agricultural mechanisation raised food-production levels. Humans were increasingly replaced by machines in the fields, which led to a huge pool of cheap labour becoming available for the burgeoning industrialisation taking root in the cities. Entrepreneurs were able to earn huge profits.

In the second half of the nineteenth century, almost a hundred years later than in England, the Netherlands saw the beginning of an industrial development that was accompanied in the cities by a dire lack of living space for the large number of workers. At that time, it was still forbidden to build outside the city walls. The result was more and more people crowding into the tiny spaces available. Some 2 million workers, 40 per cent of the country's population, lived in hovels or cellars, with their often large families in abject poverty. Furthermore, nineteenth-century towns had no amenities such as running water, sanitation systems, or regular waste-collection. It was to be some time before the Industrial Revolution began to benefit everyone, living conditions for the general population improved, and life expectancy was able to rise.

It is sobering to consider how differently this development has progressed in different countries. Life expectancy began to rise in some countries more than a century ago; other countries have only now begun taking their first tentative steps in this direction. The more tangible and intangible resources a country can generate, the higher its average life expectancy is. Thus, there is a positive correlation between gross national income and life expectancy.

WHAT WE USED TO DIE OF

Many of the things we take for granted, such as uncontaminated drinking water and clean hands, were the subject of hefty public debate in the nineteenth century. The medical establishment was seriously divided. Many believed that diseases such as cholera and childbed fever were caused by 'miasmas' — airborne substances emanating from rotting organic matter. This explanation of the origin of disease was a continuation of mediaeval medical theory. On the other side of the debate were the 'social hygienists', medical pioneers from the middle of the nineteenth century who set up local initiatives to combat widespread disease. They were convinced that the origins of disease lay elsewhere. Bacteria had not yet been discovered, theories of infectious disease had not yet been developed, and contaminated water was still an undefined concept. These committed doctors set themselves the task of promoting public health. At that time, factors such as sanitation, clean drinking water, and good food and working conditions were previously unknown innovations.

Another innovation was the use of statistical methods. Systematically recording every case of a disease in a municipality or county, and then expressing that number as a percentage of the total number of inhabitants, provides an overview of the health of the population. The mathematical tools necessary for this originated in England, where the work of a doctor called John Snow was far ahead of its time. Enumerating the sick and healthy population — the epidemiological method — was an important source of data. During a cholera epidemic in London in 1854, Snow carefully plotted new cases of the disease on maps of the city, and interviewed local residents about their daily habits, such as which public pumps they fetched their water from. In this way, he established a plausible link between cholera and contaminated water. On the basis of these statistics, practical measures were introduced to contain the disease — for example, by disabling water pumps or filling in ditches. In 1883, Robert Koch described the cholera bacterium for the first time, and it was now clear how the disease was caused and how it could be combatted. Despite this advance, Hamburg was hit by a huge cholera epidemic in 1892, which claimed some 8,600 lives. Soon afterwards, the city began in earnest with work to build a sewage and clean drinking-water system.

Cholera was not the only illness that carried people off in waves; epidemics of other infectious diseases, such as typhus, polio, diphtheria, and whooping cough, were common. And then there was 'consumption', tuberculosis, which can cause adults to waste away. The number of pathogens is endless, and they vary greatly depending on time and environment. Still today, malaria dominates the lives of people in the tropics, and influenza rages round the world every season. The flu virus afflicts the weaker members of society: children and older people. Each year, the severity of the flu epidemic can be seen from the sudden increase in the number of deaths among old people — in some winters, more die than in other years. While Spanish flu killed many people around the world in 1918–19, today bird flu takes us by surprise. New variants constantly pass from birds to humans, and so we remain susceptible targets in a hostile environment full of pathogens.

Thus it should come as no surprise that the average life expectancy in a country that had yet to see large-scale industrialisation, and where death due to infectious disease was common, did not extend far beyond 40. This number is easily explained. Half the newborn babies died early in life, and so hardly contributed to the total number of years lived. The other half of babies died in a wide range of ages, sometimes at the — for the time — extremely old age of more than 80. So the average number of years lived was around 40. There were sufficient people reaching adulthood and managing to keep their children alive until reproductive age to sustain the population.

Infections were not the only cause of early death. In many places, the climate was too hot or too cold, too dry or too wet, or a combination of those things. Whole civilisations have perished due to extreme climatic conditions. Our distant ancestors sought out sun and water to enable them to survive; but at the same time, these elements could also be the cause of disasters. In 3000 BC, the River Nile provided the basis for a rich civilisation in the middle of a desert. People faced alternate periods of drought and flooding. It was many years before they learned to live more easily with these climatic conditions — for example, by irrigating land during great droughts, and draining it when it was too wet. Eventually, this increasing control over nature led to a steady supply of food, and a dynasty of pharaohs could develop. But it is only thanks to much more advanced technological innovations that we are able — albeit with great difficulty — to settle in desert or polar regions, or in low-lying areas like the Rhine Delta. Despite modern technology, extreme heat in the summer still causes a wave of mortality among vulnerable older people. They succumb to the extra pressure that a heatwave puts on their hearts and arteries.

Even after humans abandoned the hunter-gatherer lifestyle and adopted an agricultural way of life, periods of scarcity were still a problem. Crops could fail. In addition, monoculture and an unbalanced diet were the source of major health risks, since getting sufficient calories is not enough. Vitamins, for example, cannot be produced by the human body, and have to be taken in via our diet. It was a long time before it was realised that seamen need to eat citrus fruit on long voyages to stop them from developing scurvy. From our modern perspective it seems almost inconceivable that people in the past did not understand the need to eat fresh fruit and vegetables to get enough vitamin C.

Beriberi is a disease affecting the heart that is caused by a lack of vitamin B1. The disease had been a major problem for centuries, especially in Asia. Its cause was discovered in the late nineteenth century. Dutch doctors Christiaan Eijkman and Adolphe Vorderman realised that there was something in unpolished rice that could prevent beriberi. That new insight earned Eijkman the Nobel Prize for medicine in 1929. He shared the award with the British biochemist Frederick Hopkins, who discovered vitamins A and D.

It was not only knowledge of vitamins that was lacking; little was known about the importance of trace elements. It was not until 1942 that it became obligatory in the Netherlands to add iodine to table salt, to ensure that people got enough of the element. Iodine is derived from minerals, and is often present in insufficient quantities in our food. This can lead to an underactive thyroid gland and intellectual disabilities. There are many places in the world where these symptoms are still common because the local diet provides insufficient iodine, and no action has been taken.

Alongside infections, extreme physical conditions, and inadequate nutrition, many different kinds of violence were also common causes of death among the general population in the past. As in most parts of the world, Europe's history is full of warring tribes and nations. Ethnic violence has existed since time immemorial, and the recent eruptions in the Middle East show how much impact it has on life expectancy. Part of the explanation for *why* humans are so violent lies in behaviours that have evolved via natural selection. Among social species, the alpha male is the one with the highest status. He is the head of the family, has the first right of access to sex, and therefore has the greatest evolutionary fitness. Certain personality traits are necessary to achieve such a high social status. They include dominant behaviour, which may or may not be backed up by violence. Such behavioural traits are also encoded in our DNA.

The impact of violence on the general public is often underestimated, even by statisticians. Violence can lead to the wholesale breakdown of the social infrastructure, and when that happens, statistics are no longer available. When the figures for the progress of life expectancy are drawn up, these catastrophes are euphemistically circumscribed as 'missing values' rather than labelled as extreme downward outliers. It is evident that the life expectancy of normal citizens falls drastically in times of war, whether due to ethnic cleansing or other causes. In the Netherlands, it was not the direct impacts of violence that caused a huge increase in mortality towards the end of the Second World War, but the disruption of society. The total collapse of public infrastructure during the Winter Famine of 1944–1945 led to the many deaths from starvation, cold, and infectious disease.

To provide for the basic needs of its population, and to guarantee security and stability, a country needs to produce enough goods and services. The industrial development necessary for this also results in the production of intangible resources, such as knowledge and culture. Once a country generates a high-enough gross national product to enable it to deal with the major dangers to life, making more money no longer translates into a further decline in mortality rates. Thus, life expectancy in the world's richest countries is sometimes not much higher than that in countries with average gross national incomes. It is interesting to note that babies born in two countries like Cuba and the United States, which are geographically close but whose gross national products diverge massively, now have similar life expectancies.

Good public governance is another important factor for a high life expectancy. When a government loses control of a country, life expectancy falls. For example, when the Soviet Union broke up in the early nineties, life expectancy there dropped. Excessive drinking, violence, and suicide meant that the life expectancy of men in Russia fell from 65 to 58 in a very short time. It is not difficult to imagine what effect the disintegration of states in the Middle East is having on the life expectancy of citizens there. However, the opposite can also occur: following German reunification, life expectancy rose in the former East Germany because access to public services such as pensions and healthcare improved.

In his book *The Better Angels of Our Nature,* Steven Pinker presents his hypothesis that violence in the world has declined sharply through the ages. Using comparative studies, anthropologists have established that people were nine times more likely to die a violent death in prehistoric times than they are now. Examinations of ancient skeletons have revealed that signs of violence were far more common than would be expected based on current statistics. Pinker believes that social and institutional change has led to a large reduction in the likelihood of encountering life-threatening violence. The rise of the nation-state has greatly reduced the number of tribal conflicts. To a certain extent, international trade, especially mutually beneficial exchange, has prevented governments from engaging in excessive actions. Also, 'courtly' manners have spread, literacy has increased, and democracy has become a common form of government. All this requires people to exercise more self-control and empathy. Pinker spoke at the commemorative event marking 300 years since the Peace of Utrecht, a series of treaties in 1713 that brought to an end a long period of religious wars in Europe. In an interview, he said, 'I present my argument using graphs, because most people do not believe that violence has declined. They pay attention to the events they see in the news, or point to famous wars from recent history, but they don't have the figures in their head. Once they see a graph with a line going from the top left to the bottom right, they are willing to believe me.'

THE NEW KILLERS

The recent history of developed countries shows that the typical life trajectory has changed considerably. Reaching old age used to be an exception, reserved for those few who managed to safely negotiate the hazards of life, but it has now become the rule. The ageing process has now become tangible and visible for almost everyone, and we all appear to exhibit the same symptoms. We live far beyond the 50 years in which our bodies are safeguarded from serious decay by sufficient maintenance and repair. After the age of 50, chronic illness and frailty begin to appear. The disappearance of fatal infectious diseases and the appearance of chronic illnesses and frailty is called the 'epidemiological transition' — a change in the causes of death.

Many people believe that we now die of 'new diseases', but most such diseases are not new at all. They were not recognised or defined as such in the past, for the simple reason that very few people reached the age at which they appear. There has been an explosion in the number of patients with cardiovascular disease, caused by atherosclerosis, the thickening and blocking of the blood vessels — sometimes called 'hardening of the arteries'. This epidemic is ascribed to the increase in affluence, but a more careful analysis shows that there is no clear, unambiguous connection between prosperity and atherosclerosis. The prevailing view is that the affluence we live in leads us to eat too much, smoke, and take too little exercise, and this is what makes us get sick. Nowadays, however, people from the highest social class — those who enjoy the most prosperity — have adopted a healthy lifestyle. Most people who suffer from atherosclerosis are from lower socio-economic classes. The epidemic has arisen because we are now living long enough for this disease to develop. Age is the most significant risk factor, and many older people who are not overweight, have never smoked, and who take regular exercise eventually develop abnormalities of the heart and vascular system.

There are very strong indications that atherosclerosis is *not* new and is not caused by modern prosperity. The indications come in the form of data collected from the mummified bodies of people who died in the distant past, which can now be examined using modern X-ray techniques. This shows that men and women who survived to middle age showed signs of atherosclerosis. Doctors long thought that atherosclerosis was purely and simply a result of a bad diet causing cholesterol to be deposited on the walls of the arteries. Now, the prevailing opinion is that damage to the artery walls is predominantly caused by a long-term, chronic inflammatory process. Cells produced by the immune system pass from the bloodstream into the blood vessel through the inner lining of the artery wall. Once there, they are stimulated by cholesterol deposits, causing an inflammatory response. This reaction causes damage on the spot, as if the body were reacting to an infectious alien intruder that has to be destroyed at any cost. This causes supple, easily passable arteries to turn into narrow, hardened blood vessels.

The long-term inflammation of the artery walls can be seen as a delayed, unwelcome side effect of having an immune system. This is why age is such a major risk factor for atherosclerosis. There is nothing wrong with having powerful defences that can directly do away with external pathogens. But if the system reacts too violently, or if the inflammation persists for too long, or if a person's own body is attacked although no pathogen is present, the disadvantages of the immune system begin to have the upper hand. One could argue that less inflammation is more. However, that picture looks different when seen through an evolutionary lens. If, as a pregnant woman, you need to be able to fend off pathogens in order to survive, but you also need to carry a developing baby to full term, the immune system has to be able to find a balance. The fact that this result might lead to atherosclerosis later in life is not important at the time. So older people suffering from atherosclerosis are simply manifesting the unintended side effects of having an effective immune system that can ward off childbed fever, tuberculosis, and other infections.

Biologists call this effect 'antagonistic pleiotropy'. This is an important explanatory model in the theory of ageing. It means that a trait acquired through natural selection can have more than one, very different, consequence — pleiotropy — for an individual. One or more of them can have a detrimental — antagonistic — effect on survival in old age. So atherosclerosis is primarily an expression of the ageing process. When increased prosperity results in a lower risk of dying of an infectious disease early in life, there is an — indirect — increase in the risk of dying of cardiovascular disease in old age.

Antagonistic pleiotropy provides a good explanation for patterns of illness. For example, patients with chronic bowel inflammation due to Crohn's disease or ulcerative colitis *also* have an increased risk of atherosclerosis. Similarly, a chronic inflammation of the gums — periodontitis — is also associated with an increased risk of atherosclerosis. A plausible evolutionary explanation for this is that patients with Crohn's disease, ulcerative colitis, or periodontitis have a stronger immune system than others, which leads to the build-up of a general inflammatory response to bacteria, and that also appears to target the artery walls. An alternative explanation is that such people carry a certain type of bacteria which provokes these inflammatory responses. The immune system can go even more seriously wrong. Patients with rheumatoid arthritis develop an inflammatory reaction when *no* pathogens are present. Over the course of years, that leads to the destruction of their joints, and to atherosclerosis.

All biological body and brain systems are primarily geared towards reproduction and early survival. Both our immune system and our body's energy-management system have evolved to achieve this goal. James Neel, an American researcher, has described humans as the result of natural selection in favour of a thrifty use of energy. When food is scarce, those who can survive on fewer calories — whose bodies are 'thrifty' with energy — have a survival advantage. As a consequence, we have evolved over a long period of time into people who are constantly on the lookout for food, and who can use it efficiently. When food is scarce you have to take what you can get. This is also the reason we find food so irresistible and why many of us, now living in times of overabundance, have a tendency to stuff ourselves.

Selection in favour of efficiency is, alongside our craving for food, the second explanation why we get fat so quickly, and why losing weight is so darned difficult. This is why, in the developed world, obesity and the associated increased risk of diabetes is the real price we have to pay for the cheap food we can now buy on practically every street corner.

Obesity is not just a problem in developed countries. Developing nations such as Ghana are also challenged by the problems of oversupply. Well-to-do Ghanaians have migrated to the capital, Accra. They do not have to deal with scarcity, but their bodies are still biologically designed for thrift. This is a fatal combination, and the percentage of inhabitants in the capital who are overweight is rising rapidly. In developing countries, obesity is seen as an expression of status — indicating an ability to pay for all that food. Such countries are fighting a battle on two fronts. They are trying to combat infant mortality in as-yet undeveloped areas where hunger and scarcity are common. At the same time, there is an explosion of obesity, diabetes, and cardiovascular disease among the increasing numbers of old people in the cities, where industrial development has taken hold. The causes of death in developing countries are now also no longer dominated by scarcity and infectious diseases, but by illnesses that result from overabundance, and old age. The healthcare systems of these countries are therefore in need of rapid reform.

A REVOLUTION IN MEDICAL TECHNOLOGY

Since the fifties, we have known that atherosclerosis takes a long time to develop. Autopsies performed on soldiers killed in the Korean War revealed that around three-quarters of them showed the first signs of atherosclerosis. This came as a surprise at the time, as the most striking expression of the disease — heart attacks in men — does not usually occur before the age of 50. A similar investigation of servicemen who died in the Vietnam War in the seventies showed that half of them exhibited early indications of atherosclerosis. Finally, initial signs of atherosclerosis were identified in only a quarter of those who fell in the Iraq War in the early two-thousands.

Despite its gruesome background, the picture that emerges from this data is very encouraging. It seems that atherosclerosis is now less widespread in Western populations than it used to be, and that our hearts and blood vessels are in better shape. Is this because we smoke less, and, if so, should we do more to discourage the habit? Is it because of better nutrition, and, if so, what *is* the best diet? Or are there other factors we could influence that are causing atherosclerosis to appear later in life, or not at all? Nobody knows for sure, but there can be no doubt that prevention — avoiding risk factors — has contributed to an improvement in the health of our hearts and blood vessels.

It is mainly due to prevention measures that atherosclerosis is now less common, and it is mainly thanks to interventions using medical technology that fatalities due to cardiovascular disease are on the decline. A blood vessel that has been narrowed by atherosclerosis can suddenly become blocked. A blood clot develops, which can cause a heart attack or stroke. The blocked blood vessel must be re-opened quickly with anticoagulants to prevent lasting damage due to oxygen starvation. Not even fifty years ago, doctors could do little to treat heart-attack patients other than administer morphine to alleviate the severe chest pain it causes. Then all they could do was wait. Many patients died as their heartbeat became too irregular, before their heart finally stopped pumping altogether.

The sixties and seventies saw the advent of the first 'coronary-care units' in hospitals, where patients can be sent for observation as soon as they begin to show signs of an impending heart attack, when the risk of cardiac dysrhythmia is greatest. Doctors describe the twitching motion of the heart muscles as 'fibrillation', and the problem can be corrected by giving the patient an electric shock. When this life-saving technique was first introduced, it was seen as a heroic feat. Now, we think of it as the most normal thing in the world, and defibrillators can be found attached to the wall in public spaces such as stores and stations. Once attached to the body, these devices can 'feel' what the problem is and 'know' when to administer the electric shock. Today, tiny defibrillators have been developed that can be implanted beneath a patient's skin if they suffer from repeated and unpredictable cardiac dysrhythmia.

Technology has progressed even further in the past twenty-five years. After all, why should we wait until a blocked artery causes some of the tissue of the heart to die before intervening? Today, ambulances no longer take patients to the coronary-care unit, but to the 'cath lab' — the catheterisation laboratory. There, the necessary instruments are introduced into the coronary arteries of the heart via long tubes inserted through the arteries in the groin. In most cases, a clot can be removed in this way, the narrowed artery can be dilated, and blood circulation can be restored. For most heart-attack patients who are taken to hospital, the risk of death has now been reduced to just a few per cent.

And so, in the space of just a few decades, developments in medical technology have led to a spectacular improvement in the prognosis of heart-attack patients. In the Netherlands, the statistical chance of dying of a heart attack in middle age has fallen by 80 to 90 per cent, and the situation is similar in other developed countries. This means that the condition which was so widespread among men in the developed world from 1960 to 1980 is now a thing of the past. It is interesting, although not surprising, that both men and women benefit from these developments far into old age. In the same space of time, the risk of dying of a heart attack at the age of 85 has been halved. Policymakers would like to know precisely what has led to this incredible success, but it is impossible to pinpoint which improvement is down to which innovation in medical technology. They range from health-education measures about the dangers of smoking, dietary salt intake, a sedentary lifestyle, and eating fat, to reducing risk factors with medication, acute medical care, and rehabilitation.

But the fact remains that the overall package of combined measures, in which we have invested so much time and so many resources, is extremely effective.

Compared to the acute blockage of blood vessels — that is, heart attacks among middle-aged men — the gradual hardening and narrowing of our blood vessels has received much less medical and public attention. This might be because the symptoms do not appear suddenly as an emergency that is dealt with by calling an ambulance. And it may also be because the symptoms of atherosclerosis mainly appear in old age. They are seen as simply part and parcel of being old, and we consider them 'normal'. However, that is no reason not to do anything about them; such an attitude would be a waste of much of the healthcare benefit achieved up until now. Just as the acute blockage of blood vessels leads to tissue damage, a gradual narrowing leads to reduced blood flow, a chronic shortage of oxygen, an accumulation of different kinds of minor damage, and a gradual decline in the function of the heart, the kidneys, and the brain.

We need a new revolution in medicine and technology — not, this time, to enable us to deal with acute problems, but to help us stop the gradual hardening and narrowing of our blood vessels. People who have developed very little atherosclerosis during their lives experience a less-rapid decline in heart, brain, and kidney function in old age. This means their organs age less quickly. The risk factors for the sudden and the gradually developing problems are mostly the same, but high blood pressure is a major contributor to the hardening and narrowing of blood vessels in old age. Given that only a quarter of patients with high blood pressure are identified and treated, there is still a world of progress to be made.

In the second half of the twentieth century, half the population died of cardiovascular disease. Now, at the beginning of the twenty-first century, the proportion of the population that succumbs to this cause of death has dropped to a third, and is expected to continue to fall. Other causes of death, especially cancer, are increasing, and have taken over the top spot. Combatting the causes of death is a bit like peeling an onion: every time you remove one layer, another appears. Currently, one person in three dies of cancer. Impressive results have been achieved in the treatment of specific kinds of cancer. For example, the use of bone-marrow transplants has led to a large drop in the number of deaths from leukaemia. Other kinds of blood cancer can also be treated successfully with a combination of radiation and chemotherapy. However, a general fall in the significance of cancer as a cause of death, as we have seen with cardiovascular disease, is not in sight. To begin with, there is expected to be an epidemic of lung cancer among women, since they took up smoking in large numbers in the sixties and, unlike men, have not yet started to kick the habit.

AN EXTRA WEEKEND EVERY WEEK

In 2002, the leading academic journal *Science* published a paper on the change in human life expectancy since 1600. A graph showed which country was at the top in any given year. From 1600 to 1800, precious little happened, and life expectancy at birth fluctuated around the 40-year mark. Occasionally, it dipped below 40, when famine, the climate, or war took their toll. Things changed around the year 1800. Life expectancy in England began to rise steadily. Whenever the Industrial Revolution took hold in a country, financial resources became available for investments, society became more organised, and a rise in life expectancy followed. One country after another took over the position of frontrunner. After England's start, the list was topped for a long time by the Scandinavian countries, followed later by New Zealand. Between 1940 and 1970, the Netherlands was regularly top of the life-expectancy charts — sometimes for men, but more often for women. After that, the Netherlands fell back to about the middle of the field. For the past twenty-five years, Japan has topped the list. There, life expectancy has now risen to 80 years for men and 87 for women.

The most remarkable thing about these records is that they form a straight line — the average life expectancy in developed countries rises by two to three years every ten calendar years. This may not sound like much, but that rate of increase can be expressed another way: every week we get an extra weekend of life expectancy; or, every day we gain six hours. Within the space of one century, human life expectancy has doubled, and some countries will reach the 80-year mark in a much shorter time. China, Chile — will they be the new champions?

In the past, various private and public institutions — such as banks, insurance companies, and the World Health Organisation — have made predictions about the increase in average life expectancy at birth. In 1920, it was predicted that average life expectancy would never exceed 65 years. In fact, at that very moment, life expectancy in New Zealand had already topped 65. Again and again, actuaries came up with new estimates of the highest possible age. And each time, that record was soon broken somewhere in the world. Actuaries made these pessimistic prognoses because they assumed there had to be a biologically determined maximum possible age for humans. But according to modern insights into evolutionary biology, (human) life knows no such limits. Whenever we increase our investment in the preventive care, maintenance, and repair of our bodies and our brains, we limit the accumulation of lasting damage, enabling us to stay healthy and to live longer. The numbers also show that the concept of a maximum possible human age is false, since time and again the assumed maximum has been exceeded. The current record is already 122 for women and 116 for men. And it is only a matter of time before those records are broken, too. We must get away from the idea of a maximum possible length for human life.

It used to be thought that life expectancy could only rise thanks to a drop in infant mortality. And that was the case at first. If all children stay alive, and infections such as tuberculosis are consigned to the history books but nothing is done to combat the ageing process, the average life expectancy rises to around 65 to 70 years. But in the 1950s, the statistical chance of dying in old or very old age started to fall drastically in the leading countries on the list. This explains why life expectancy continues to rise, and why it does so at the amazing pace of two to three years of life every ten calendar years. The fall in the chance of dying at an advanced age is largely due to new insights into the treatment of cardiovascular disease. First, the number of deaths due to heart attacks in middle age declined, and this was followed by a drop in deaths due to abdominal and peripheral (belly and leg) vascular disease in people in their 60s and 70s. Now, the number of strokes among the oldest in society is also beginning to decline. So this increase in life expectancy has not come out of thin air. It is the result of financial investment, technological development, social engineering, healthcare, and innovation.

But 'classic' causes are still often the reason why the development in life expectancy stalls or even reverses in certain countries or regions. For example, the tsunami in Fukushima temporarily ousted Japan from the top spot on the list, because this extreme event claimed so many victims. Repeated periods of famine limit life expectancy in North Korea. The outbreak of civil war in Syria in 2011 caused a fall in life expectancy there. The massive spread of HIV has prevented the average life expectancy in many African countries from reaching or crossing the 40-year mark. Natural disasters, the availability of food, social disintegration, and infectious disease continue to impact on life expectancy today, but they are no longer the dominant factors.

People often ask me how life expectancy will develop in the future. Sometimes there is an undertone of disbelief or cynicism in the question — for example, when I stated publicly that the first person to reach the age of 135 has already been born.

'The best basis for estimates about the future is the increases gained in the past,' I say. 'Add two to three years to today's average life expectancy over the next ten years.'

Then someone asks, 'And what then?'

'After that, it will increase by another two to three years,' I answer.

This last remark is often met with amazement, and sometimes irritation.

'But people can't keep on getting older forever!' another interjects angrily.

These reactions show that people often don't want to face the facts. That could be due to their own fear of getting old.

Then I make the bold statement that ageing is avoidable, and that dying is not encoded in our DNA, so we can continue living longer and longer. I usually fail to convince my interlocutors over a cup of coffee that natural selection only has a grip on the beginning of our lives, and not the end. This usually puts an end to the conversation.

So I usually don't try to convince people with the evolutionary argument, but continue the conversation by saying, 'Besides, people have often tried to estimate how high the ceiling is, but they have always turned out to be wrong. Their estimates were far too conservative. Look at the statistics: two to three extra years per decade is a realistic figure.'

If that goes down well, I step it up a notch: 'There are not just conservative and realistic estimates out there. Some projections are downright optimistic. Some of my fellow researchers in the field, like Aubrey de Grey, for example, think technology will allow us to raise life expectancy even further. If increasing medical and technological knowledge enables us to keep ahead of the accumulation of lasting damage — for example, by repairing or replacing cells, tissues, and organs — it could be that the first person to live to 1,000 is already being born about now.'

'Why don't people live forever, then?' is the next question.

Whereupon I explain, 'You can avoid ageing, but that doesn't make you immortal. There is always the danger of becoming the victim of a sudden death. For example, you might get run over while crossing the street.'

The idea that we could always remain one step ahead of lasting damage is not nonsensical. We have been replacing worn-out heart valves for many years, as well as hips, lenses clouded by cataracts, kidneys, and sometimes an entire heart. But the idea is rather optimistic. It would require much more sophisticated medical and technical interventions to become reality.

Often, the conversation then goes on: 'So, what about my daughter? She was born in 2000. What will happen in her lifetime?'

'Just do the maths: current estimates say she will live to over 80, but for every ten years she lives, she will get an extra two to three years of life expectancy. Let's round that off to a hundred. Tell your daughter she will live to 100.'

'My daughter — 100 years old?!' That reaction is typical of the incomprehension and resistance that the increase in life expectancy provokes in people.

'Of course, that's only if she doesn't go off the rails. Keep her away from the world of sex, drugs, and rock 'n' roll. These are all average figures, of course, and there are lots of ways you can screw up your life along the way.'

7

LEGIONS OF THE OLD

Even though most people today get old, old age is no longer associated with the magic and mystique it used to be. Death used to have little to do with age, but in our times, age, illness, and death form a triad. The large population of old people has great implications for the demographic structure of society. While it used to resemble a pyramid, it now looks more like a skyscraper — with a temporary bulge representing the baby-boomers. The ratio between the working population and those under 20 years of age (the young-age dependency ratio), and that between the working population and those aged over 65 (the old-age dependency ratio), changes over time, but the total dependency ratio — that is, the ratio of economically active to inactive members of society — remains roughly the same. The old-age dependency ratio can be reduced by increasing the retirement age.

It's New Year's Eve, 1889. In Lochem, a village in the east of the Netherlands, not far from the German border, the municipal gravedigger, Jan Hendrik Lenderink, is drawing up his report on the year's deaths. The next day, Lenderink has to present his list of the deceased to 'the Right Honourable Mayor and Aldermen, Councillors, and other Residents of Lochem'. That year, 85 residents died. In his report, the gravedigger categorises them according to age. There have been six stillbirths, and 27 babies died below the age of one. Another six died before they could celebrate their third birthday. Almost half of the bodies he has buried this year were those of babies, infants, and toddlers. Lenderink thinks nothing of this; for him it is routine, business as usual. And it was not much different for his father, the municipal gravedigger before him. He told Jan Hendrik of the countless young lives he saw nipped in the bud. Just two days, two weeks, or six months of life — and then, all of a sudden, dead. Their parents were apparently unmoved by this state of affairs. Death was a part of life. If you had told father Lenderink and his son that parents in Africa do not name their children until their first birthday, they would have understood why.

THE GRAVEDIGGER

A municipal gravedigger was responsible for making sure the burial of the dead went without a hitch. In his shed, he always kept a stack of small coffins for the little bodies of dead children. They were buried with a simple but solemn service in a special section of the graveyard. How different this was from the great care and ceremony that went into the burial of youths, adults, and old people. Their bodies were first measured at their homes by Lenderink, so that he could build a coffin of the right size in his workshop. Young people died of polio; men perished from consumption, or were stabbed to death in an argument that got out of hand; proud women who had just become mothers were taken soon after childbirth. There was no predicting it. Of course, there were also men and women who died of cancer or dropsy; today, we would diagnose cases of dropsy as heart failure, but at the time it was not recognised as such. Finally, there was the rare individual, shrunken and senile, a phantom from the past, who slipped away quietly in her sleep. So, in his time, Lenderink experienced death in all age categories. Most of the people he buried had met an early end.

In the nineteen-thirties, unrest began to spread among gravediggers. Yes, it was during the Great Depression, but that was not it. The profession seemed to be changing. Some gravediggers spoke of a strange trend. They noticed that fewer and fewer children were dying, while the number of old people they were burying was increasing all the time. A veritable industrialisation of the coffin-making business seemed to have begun. One gravedigger announced that keeping a stock of these enormous coffins was now standard practice for him. Another told how a colleague had opened a shop to display the different kinds of casket that were now available. Some gravediggers were not the least bit surprised to see the children of the deceased perusing the range in the shop, choosing their parent's coffin. But the rest listened in horror. They did not recognise any trend in this, and believed it was pure coincidence that more people were dying at a greater age. They were sure the situation would soon return to normal. They were proud of the fact that for generations they had provided simple coffins for the little ones and personal, bespoke caskets for the adults. They saw no need to change the way they did business, just because 'getting old happens to be in fashion'.

Nobody in the village knows exactly when Lenderink the gravedigger gave up working. Was it just before the war, or just after? Anyway, it was inevitable, say the funeral directors who now organise burials in Lochem. Lenderink continued to build coffins to order, but even that ceased to be profitable at some point. Today, a carpentry workshop is no longer necessary. A showroom, however, is. Factory-made coffins come in all sorts of sizes and designs. Burying little children, once routine for a gravedigger, has become a rarity. These days, no one really thinks about the possibility of a child dying. And when it does happen, family and friends find it almost impossible to come to terms with. Who, these days, expects a child to die? A lot of care has to be taken in finding an appropriate way for the dead child's loved ones to say goodbye. On the other hand, it has now become common for old people to arrange their own funerals.

Now, the risk of dying before the age of 50 is small. Almost everybody these days dies later than that, though almost everyone dies before the age of 100. The average age of death now stands at around 85 years. Death has been tamed, from an ageless monster to an almost predictable fate for older people. In 1889, no one could be sure whether they would still be alive in a year's time. Under the conditions at the time, old age was a desirable commodity, and old people provided a glimpse of the mystery of holding on to life, long into old age. Nowadays, so many people reach old age that it has lost the magic and mystique of yesteryear. Gone are the days when a grey beard was associated with wisdom and valuable life experience. Today, old age, illness, and death form a close triad. There is little interest in the lessons learned by old people over their long life, or in their capacity for reflection and their powers of empathy.

FROM PYRAMID TO SKYSCRAPER

To understand the consequences of this new lifespan, it is a good idea to imagine what life was like around the year 1900 — before this revolution took place. That's no easy task. Of course, we have paintings and photos from that era, but they provide a distorted picture. They are mostly images of celebrations and parties, made for people with a lot of money, as a way of capturing highpoints in the life of their families. But images of ordinary life, of the normal scene on the streets, are much rarer. I imagine it to look a little bit like the markets you still find in north-eastern Ghana today. It is a crowded scene; everybody has come in from the surrounding villages to sell their wares and to buy things themselves. Children swarm about, some of the older ones with a toddler on their back. There are crowds of youths standing round carts, mothers with babes-in-arms, men selling cloth. And here and there, sitting in the corner, you can see a single, toothless old person.

This is nothing like the scene at the street markets in our modern cities. There are droves of older people, out doing their shopping. They are grey-haired, but their mouths are full of shiny white teeth. Middle-aged couples stroll past the market stalls, pushing their grandchildren along in buggies. A few young fathers and mothers do some last-minute shopping in a rush. Very occasionally, you see a group of youths hanging about.

This common street scene reflects not only our affluence and culture, but also the structure of our population. The market scene in north-eastern Ghana is a typical example of the population pyramid. Such a structure develops whenever death is able to strike mercilessly and unchecked right from birth. As the years pass, almost no one is left alive. But the West has made short work of that. Soon after the end of the Second World War, the World Health Organisation was set up to tackle major medical and social problems. In the second half of the twentieth century, a large number of 'mother and child care' programmes were introduced — all sorts of initiatives to reduce mortality associated with childbirth, to give newborn babies a realistic chance of a future in regionswhere that was not already the case, as well as sanitation and hygiene programmes, obstetric care, nutritional advice, and vaccination drives. Where these programmes have been successful, and child mortality is reduced to a minimum, and death takes its toll only in old age, the demographic pyramid takes on the form of a skyscraper. Generation upon generation, storey upon storey, are filled with people, (almost) all of whom will stay alive and reach an ever-increasing age. This change from pyramid to skyscraper is called the 'demographic transition' — that is, the change in the age structure of the population.

The demographic transition is more than just a change in the composition of the population. Before the transition took place, women commonly had six to seven children, of whom two to three reached adulthood. When mortality figures fall, and babies, infants, toddlers, and young adults stay alive, the population increases in size. The number of births surpasses the number of deaths, in what we call a 'birth surplus'. This phenomenon occurred in the twentieth century in almost every country. It explains how, from 1900 to 2010, the number of people in the world exploded from 1.5 to 7 billion.

People have been worrying about the rapid increase in the global population for a long time now. As early as 1972, the Club of Rome reported that the situation was 'unsustainable'. The paper predicted that we would not be able to feed the ever-increasing number of mouths. At that time, the world population stood at around 4 billion, and a veritable Malthusian scenario was expected. This prompted a call for intervention policies, and massive birth-control programmes were instigated. Such programmes had very varied rates of success. One of the few countries that managed to curb its birth surplus early was China, with its one-child policy. But despite the explosive growth in the world's population, mass starvation has not occurred. Just like Malthus in his time, the Club of Rome hugely underestimated our capacity for innovation. At the beginning of the twenty-first century, we produce enough food for approximately 12 billion people. About one-third of that goes to waste. Today, starvation is principally a problem of distribution. Political conflicts, economic decisions, and acts of war are responsible for people not getting enough calories and/or nutrients.

It is noteworthy that family size decreases naturally when a country goes through a demographic transition. Parents simply start having fewer children. We saw this happen in our study area in north-eastern Ghana, where a large fall in mortality rates was accompanied by a parallel reduction in birth rates. When we began our study there in 2003, women gave birth to an average of six children each. Ten years later, that number had fallen to two or three, and there was no interventionist population policy behind that development. It was as if parents sensed that the circumstances in which they were living and starting families had improved greatly. They no longer had to invest in having six children, since they felt that two or three were now enough to ensure the next generation.

The reproductive behaviour of parents in the Netherlands was no different. In 1900, women gave birth to more than four children on average. That number decreased steadily to a level of 2.5 in the 1940s. After 1968, a sharp decline began, and the average number of children per woman reached 1.5. The reasons for this included not only the introduction of the contraceptive pill, but also increasing prosperity, emancipation, secularisation, and individualisation of society. This was also the period when the standard life trajectory — marry early and have children — gave way to a more flexible pattern. Change in the composition of the population due to a fall in mortality is called the 'first demographic transition'; the population change due to rearranging our family lives is known as the 'second demographic transition'. The number of children per family in the Netherlands has now risen slightly, to 1.8, but that is still well below the minimum of 2.1 necessary to maintain a stable population over time. The minimum number is larger than 2.0 because a small proportion of each generation will die before adulthood, or will be infertile.

The population explosion in the twentieth century

has its origins particularly in the imbalance between the number of births and the number of deaths. However, since the number of births is currently falling in almost every country, global population growth is now levelling off.

Almost every country has to deal with a baby boom that originated from a period when births and deaths were temporarily not in equilibrium. This explains why the population makeup in many nations does not take the straight-sided form of a skyscraper, but has a bulge somewhere in it. A prime example is the wave of births after World War II. Especially in developed countries, the number of newborns surged, only to decline more than a decade later. The Dutch saw a huge wave of births. The United Kingdom faced a short baby boom immediately after the war, peaking in 1946, and a second baby boom with a peak in births in 1964. In the US and Australia, servicemen and servicewomen returned to their homes and resumed family life after several years of wartime conflict. In all cases, this had a major impact on the population structure, the imprint clearly being visible in the population 'pyramid'. These baby boomers are now ageing, leaving the labour force and entering their senior years. This is a frightening prospect for economists and politicians alike. It will make the populations grow older, with a peak set to occur around 2040. After that, however, the populations will become younger again.

YOUNG AND OLD-AGE DEPENDENCY RATIOS

We can see a number of socio-political consequences of changes in the structure of the population. If it is shaped like a pyramid, a relatively large proportion of public finances must be invested in children. They need to be brought up and educated. Whenever there is a wave of births, the consequences for society are easy to predict. It is no coincidence that the sixties and seventies were a time of many important educational reforms in the Netherlands. The 1968 Education Act was introduced as an attempt to provide the growing numbers of young people with the best education possible. The status of middle and secondary modern schools was raised, and they were merged with classical grammar schools to create large school complexes. Existing school buildings were surrounded by countless temporary structures to house the large numbers of pupils. Not everyone in the Netherlands has fond memories of those huge schools, sometimes with two to three thousand students.

Demographically speaking, there was an enormously high 'young-age dependency ratio'. This is a statistical measure expressing the ratio between the number of people below the age of 20 and the size of the working population. In the Netherlands, the latter group is officially defined by Statistics Netherlands as people between 20 and 65 years of age. From 1960 to 1970, the young-age dependency ratio in the Netherlands stood at around 70 per cent. This means that every ten working people were socially responsible for raising seven newborn babies to adulthood — no easy task.

The decline in the number of births means that this pressure has decreased; in 2010, the young-age dependency ratio in the Netherlands stood at 40 per cent. This means there were ten working people to organise and finance the upbringing and schooling of every four newborn babies.

This has allowed the scaling back of investments in mother and child welfare, vaccination programmes, schools, and other public- and private-sector investments.

Now that our responsibilities for young lives have become less, the increase in the number of older people poses new challenges. There is an increasing demand for care from older people who suffer from chronic illnesses, have become frail, and require help, as a result of the ageing process. In other words, the old-age dependency ratio is increasing.

Some politicians and policymakers, but also ordinary citizens, are seriously concerned about the increasing number of older people who will become dependent on the working population. The old-age dependency ratio in the Netherlands is set to double from 20 to 40 per cent by 2025, and to rise to 50 per cent by 2040, at the height of the ageing population phenomenon. In other words, while every ten workers used to have to bear the public responsibility for two older persons, that number will rise to four or five in the future. There is much debate about whether that represents a large number, and many people wonder whether it is even possible to manage an old-age dependency ratio of that size at all.

A glance at the situation in other countries helps put the situation in the Netherlands in perspective. In Japan, the old-age dependency ratio has already risen to between 40 and 50 per cent; and in Europe, countries such as Italy and Germany have moved much further in that direction than the Netherlands. What is more, the birth rate in those countries is even lower than in the Netherlands, and their populations are shrinking. The population in the Netherlands is still growing slightly due to immigration, and that also curbs the increase in the old-age dependency ratio. China's profile is almost identical to that of the Netherlands, which is unexpected, given its one-child policy and the sharp drop in mortality as a result of the country's rapid economic development. The explanation is that China will be able to fall back on large numbers of workers for a long time to come, since its strict family-planning policy was only introduced in 1979. If we look at the United States, we see that there is no increase in the old-age dependency ratio on the horizon. Although women in the US are also having fewer than two children on average, the country has a huge number of immigrants who will keep the working population at the necessary level.

Fear of the consequences of this development in the old-age dependency ratio has fuelled recent calls for interventionist population policies. These calls are principally based on advocating ways to persuade young parents to have more children. Apart from the moral question about whether it is the place of governments to interfere with family planning, it is very questionable whether it would even be of help in the long term. First of all, the young-age dependency ratio would rise. The average cost to parents and society of caring for a child from birth to employment age in Europe is estimated around 200,000 euros, but estimates vary widely between countries and socio-economic class. Researchers from the University of Canberra have found that it costs a middle-income family in Australia as much as $812,000 to raise two children, while the estimates are $474,000 for lower-income families and $1,097,000 for higher-income families. The greatest expenses are for food, transport, and education; however, the latter very much depends on whether children are sent to private or predominantly state-funded schools.

In the years that follow, the individual has to pay back that 'debt' to society. A person only begins to profit society from the age of 40. The fact that it takes 40 years before a person is 'quits' with society is in accordance with the fact that the minimum average life expectancy necessary for a stable population size is 40 years. This also means that all the years that we continue to benefit society, both tangibly and intangibly, beyond the age of 40 can be seen as that part of the gross national income that is free spending money. Conversely, it also explains why countries fail to 'progress' when life expectancy lags far behind, as is the case in areas heavily affected by HIV.

Another argument against the call for people to have more children is that it leads to an inevitable increase in the size of the population. This may not be an important objection in sparsely populated countries, but it is very doubtful whether it is really what the Netherlands or any developed nation should be striving for. There is general agreement that it would be better if there were fewer people living on Earth. Issues around environmental policy, renewable energy, and biodiversity are at the top of the list of 'major challenges facing society'.

The third argument is the most compelling reason to reject such demographic intervention. Every wave of births will eventually lead to a wave of old people, since children get older as time goes on. In other words, trying to relieve the pressure of the old-age dependency ratio with an intervention that will increase the old-age dependency ratio in the future results in a vicious circle, and saddles future generations with a burden.

There is one way to curb the rise in the old-age dependency ratio effectively, and that is to raise the dividing line between working and not working, between productivity and dependency, from 65 to a greater age. This increases the size of the working population, and simultaneously reduces the number of people who are dependent on others. Such a move would greatly reduce the old-age dependency ratio, and is more than just an accounting trick to deal with the greatest challenge facing society in our times. There is something foolhardy about our stubborn adherence to 65 as the pensionable age, as if it had some kind of biological basis. The number has come to be seen as a dividing line in people's lives between a period of giving and a period of taking. But, in fact, this delimitation is based on nothing but a political compromise that was reached in the nineteenth century. I return to this point in Chapter 13.

The question is whether the problem of the ageing society in developed countries really is as great as people believe. If we limit our view to the increasing pressure of the old-age dependency ratio, then the clouds on the horizon really do look black. But what people often forget is that the pressure from the young-age dependency ratio has relaxed. This means that the overall dependency ratio, between the working and non-working sections of the population, has not increased significantly. To compare: in the Netherlands, the total ratio in the sixties and seventies stood at 90 per cent, comprising 70 per cent young-age dependency and 20 per cent old-age dependency, with almost equal figures in the UK. The overall dependency ratio in Australia and the US were even higher, as their populations were relatively young when compared to the

European continent. At the end of the twentieth century, the total ratio was lower, because that was when the post-war baby-boomers were adults, actively contributing to the labour market, and had not yet grown old.

However, we expect that the overall ratio in the Netherlands will also stand at 90 per cent in 2040: 40 per cent young-age dependency and 50 per cent old-age dependency, and that it will be similar again in the UK. Overall dependency rates in Australia and the US will be in the same range, but with 10 per cent more youngsters and 10 per cent fewer elders. To put it another way: every individual will have to share the responsibility for one other person during their working life, and that is just as true of the future as it was in the past. While our principal responsibility used to be raising the young to adulthood, in the future it will mainly be older people who require our care and support. Most people prefer the idea of wiping babies' bottoms to that of wiping old people's bottoms, but that is a different matter. From an economic point of view, it is far from being a doomsday scenario.

Although the old-age dependency ratio is manageable at the national level, it can increase drastically at the local level due to migration. The population in the north and south of the Netherlands, for example, is shrinking, because young people in particular are moving away. This results in an ageing regional population, although not because birth rates are falling or because people are suddenly growing old. Local population ageing can become a serious problem without it showing up in the national statistics. There appears to be no problem on the level of the national budget, and the old-age dependency ratio can be managed financially. But this does not take into consideration the fact that the local population is increasingly out of kilter. The problem then is not a lack of money, but a lack of people to keep society running. Who will manage the local supermarket? Who will staff the library? Who will help old people with their everyday tasks, like housekeeping and washing?

The phenomenon of local population ageing is eternal and ubiquitous. The famine in nineteenth-century Ireland was caused to a large degree by the depopulation of rural areas. The Industrial Revolution attracted workers to the cities. The very young, the old, and the needy remained behind. In today's China, young adults are moving to the cities, because that is where careers are to be forged. The older generation stays behind in the countryside, often left looking after their grandchildren.

All over the world, lively communities are turning into ghost villages. Sometimes, a picturesque hamlet is bought up wholesale and turned into a hotel resort. Norway is trying to turn the tide by offering residence permits to skilled professionals from abroad, on the condition that they settle and work in a local community. The appearance and disappearance of communities is comparable to the appearance and disappearance of species. It is a never-ending cycle.

8

AGEING IS A DISEASE

Ageing and disease are inextricably linked. Partial causes of disease accumulate as we get older, until, as it were, our 'bingo card' is full, and illness strikes. Ageing is not 'normal', because it results from damage that occurs where there was originally none. If ageing were normal, it would not be necessary to investigate ways to prevent it. A common form of ageing-related damage to the brain is dementia, which covers much more than just Alzheimer's disease. All indications are that fewer people in the future will develop dementia than is the case now. But older people will continue to become frail, and that is a key insight in geriatric medicine.

On 25 July 2000, an Air France Concorde crashed. Shortly after taking off for New York, the plane developed serious problems. A few minutes later, it came down on a hotel close to Charles de Gaulle airport in Paris. All 109 people on board, as well as four staff members at the hotel, were killed.

The crashing of that supersonic aircraft led to a highly detailed criminal investigation, which eventually resulted in a final verdict in 2012. The fact that there was still debate about the cause of the crash twelve years later says a lot about the origin of this kind of accident, which involves the failure of complex systems.

During initial inspections, pieces of rubber were found on the runway. Could a tyre blowout have caused the disaster? Research into the Concorde's flight history revealed that there had often been blowouts, but they had never caused any serious problems. Initially, that did not seem to the investigators to be a likely explanation.

Eventually, they reconstructed the following concatenation of events. It began when a Continental Airlines plane took off from the runway before the Concorde, and lost a titanium-alloy strip measuring about 50 by 3 centimetres. That metal strip caused the blowout while the Concorde was racing along the runway prior to taking off a few minutes later. What had never taken place before now happened: a piece of the blown-out tyre hit the fuel tank above it and caused it to rupture. An electric spark then ignited the fuel that came gushing out of the tank. In response to the fire warning in engine 2, the crew shut it down. The explosion had caused so much damage to the electronic and hydraulic systems that the landing gear could not be retracted, and the plane failed to gain enough speed to climb further. With just three of its engines working, the aircraft banked sharply to the left, so the crew throttled back power to engines 3 and 4, causing the Concorde to lose even more speed. Then the plane fell out of the air.

So what was the cause of the disaster, and whose fault was it? Was it caused by the titanium strip, the tyre blowout, or a design fault? On 4 July 2008, the families of the victims brought a case against two technicians from Continental, the airline that owned the plane that left the titanium strip on the runway. The charge was manslaughter. The head of Aérospatiale, the company that built the Concorde, was also charged, since he was revealed to have known of more than seventy previous incidents involving Concorde's tyres. The prosecutors accused the aviation authority of failing to take appropriate steps to ensure the plane's safety. Finally, an engineer who had been involved in the design of the supersonic aircraft was also in the dock. It was claimed he knew that the design was not entirely safe. In 2012, an appeals court ruled that charges of manslaughter could not be proven against the accused.

Not surprisingly, the entire fleet of Concordes was grounded immediately after the Paris disaster. After the initial investigation had established the most likely chain of events, it was decided to develop burst-resistant tyres, and to line the interior of the wheel arches with shock-absorbent material so that the fuel tanks would no longer be susceptible to rupture. A reconstruction of the relevant part of the plane was built, and the scenario was tested experimentally, with encouraging results. The chain of events that gave rise to the disaster was effectively broken, and, after the necessary adjustments were made, the aircraft was once again given a certificate of airworthiness.

WHAT CAUSES CANCER?

Pathogenesis — that is, how disease develops — has much in common with the Concorde disaster. Especially disease that occurs in old age. There is never one single cause; it always has a series of partial causes, which together are enough to explain the onset of disease.

One example: smoking is seen as *the* cause of lung cancer. But if that is the case, why does only one smoker in every five develop the disease? Perhaps the other four died of something else before lung cancer had a chance to arise. Some people reason that if those four smokers had lived long enough, they, too, would have developed lung cancer. Others believe that some people are simply lucky and others are not. But medical science does not deal in random chance. There is always a biological mechanism behind the appearance of disease, even if we sometimes do not know what that mechanism is. As already mentioned, there is never just *one* cause, but rather a combination of various partial causes.

Lung cancer occurs when cells multiply uncontrollably at the expense of the surrounding tissue, and then spread. Those malignant cells are able to avoid the stimuli from their environment that normally keep rampant cell-growth in check. Malignancy can only raise its ugly head if there are faults in various parts of the cell's DNA that are essential for normal cell division. One single mutation in the DNA is usually not enough to cause cancer. It is the combination of specific types of damage in various genes that makes the rampant growth of malignant cells possible. This proposed mechanism of pathogenesis is called the 'multiple-hit hypothesis'. The accumulation of 'multiple hits' to the cells' DNA causes cancer to develop. In the pathogenesis of cancer, smoking alone is sufficient to damage the DNA, the genes, of a cell. Smoking is like a man with a gun trying to hit a target — the DNA — with a random hail of bullets. But that alone is not enough to explain how lung cancer occurs.

Damage to our DNA — or, more precisely, damage to the genes that contain the code for life — is just one side of the matter; the other is the ability to repair that damage. The molecules of DNA in every cell are constantly checked for errors, and, once identified, mistakes are corrected. The repair of faulty DNA molecules is therefore essential for the maintenance of our genes, which, in turn, is necessary for keeping our cells, tissues, and organs in good order.

Human evolution has selected in favour of this ability, but we are not all equally good at it. Sexual reproduction creates constant variations among our offspring. Too little DNA repair can cause problems — for example, cells lose their ability to function. But investing a lot in the maintenance of our own DNA, so that no damage at all occurs, brings no evolutionary advantage, since we are not built to last forever, and our children already carry good copies of our DNA. Furthermore, DNA repair is an intensive process that must be made at the expense of investing in other processes, such as reproduction. Thus, DNA repair illustrates the disposable soma theory translated to the molecular level. The genetically determined variation in DNA repairability can offer a good explanation of why cancers appears more often or earlier in life among members of certain families. Others may smoke like chimneys and still never develop lung cancer.

The vast majority of lung cancer patients are smokers or ex-smokers. However, there are also patients who get lung cancer although they have never smoked in their lives. There are countless other processes that can damage DNA and allow cells to grow out of control. For example, much damage is caused by free oxygen radicals, which arise when glucose is metabolised to provide cells with energy. DNA is a large, complex molecule, and damage can occur in various places and in various ways within one individual. This means that all patients have their 'own' tumour. Patients can develop several tumours in their lungs, and they are not necessarily all exactly the same. Thus, lung cancer is not one single disease. Each tumour is the result of a unique set of partial causes, which together are 'sufficient' to cause cancer. And this is no different from the way every plane crash is the result of a unique chain of partial causes.

Damaged cells must be removed from the body, to prevent cancer and other mischief. The mechanism for this is called 'apoptosis', or programmed cell death. It has been selected for in our evolution, and dismantles cells from the inside, causing them to implode, as it were. This prevents a cell from bursting and leaking its contents into the surrounding tissue, where it could cause an inflammatory reaction. Apoptosis is also the mechanism that gets rid of excess cells as we develop in the womb. A classic example of this is the way our fingers develop: cells are removed from between the bones of the budding digits, like a sculptor chipping away bits of stone, and our fingers are formed, so that we don't have to go through life with webbed hands.

When a cell has accumulated a lot of damage to its DNA, an apoptosis programme is activated inside it. A 'counting machine' also keeps track of whether the cell has already divided many times and might be worn out. The DNA in cells that have undergone many divisions may contain an accumulation of copying errors, and those cells are removed as a precaution. Every time a cell divides, the telomeres within it get a little bit shorter. Telomeres are the ends of the DNA molecule, and are, as it were, the end of the rails along which the DNA copying machine runs, like a train. When the rails become too short, no more copying can be done, and the cell stops dividing.

Because the ability of cells to divide is crucial for tissue repair, telomeres were originally thought to be a clue to cause of ageing when they were first discovered. However, this turned out not to be the case. Mice have longer telomeres than humans, but age more quickly. Also, average telomere length says nothing, or very little, about how long we will live. It is more probable that telomeres developed as a protection against cancer. Some cells that carry out special functions are 'arrested' when their telomeres have become shortened. They can then continue to exist without the risk of cancer developing. Other cells meet a much more drastic fate, and are removed from the tissue of the body.

NORMAL AGEING DOES NOT EXIST

There is no single cause that can explain the signs of human ageing. Ageing arises due to a unique combination and interaction of partial causes, which renders us increasingly infirm and eventually results in death. The combination of partial causes is different for everyone, and ageing takes a different course in each of us: sometimes it progresses quickly; sometimes slowly. An older person's kidneys might still be functioning fantastically well, but he can still be hit suddenly by a stroke.

Failing organs can be compared to old beer glasses that have accumulated tiny breakages over time. Humans, too, acquire more and more different kinds of minuscule damage as they pass through life — for example, from smoking, oxygen free radicals, an injury, or an infection. One kind of minor damage is in itself insufficient to cause an illness or infirmity. That's why you think you are healthy. But if a doctor were to examine your body extremely carefully, she would notice all the minute damage that has accumulated over time. A whole-body scan will always reveal a certain number of defects, and that number will be higher the older the patient is.

A whole-body scan that revealed *no* abnormalities would be really remarkable, but the fact that such scans are increasingly being used when there is no apparent need for them is a worrying development. Usually, there is no need to do anything about the defects they reveal. It is not always better for a person to undergo more and more diagnostic testing for damage that is only minor. Indeed, the side effects, both emotional and physical, of such a scan can leave a patient worse off.

This accumulation of minor damage is what makes people frail in old age. No one develops a disease 'just like that' or 'suddenly' falls ill. The missing link, the final partial cause that triggers a biological mechanism, is what leads to the sickness or disease. It is like a game of bingo: as the game progresses, players' cards are increasingly full of marked-off numbers, then one more number is called and BINGO! — you have a full house. In the bingo game of illness and disease, the chance of getting a full house increases the older you get, since you acquire more and more partial causes. Your body becomes frail and vulnerable.

Many researchers, doctors, and patients try desperately to make out a difference between the ageing process and the onset of illness and disease in old age. Is a worn-out knee a result of ageing, or is it an ailment or disease? In principle, there is no difference. After all, ageing arises due to an accumulation of partial causes, just like sickness and disease in old age. Looking in the mirror, we can see the wrinkles advancing and our jowls beginning to sag. The elastin that keeps our face straight is damaged, just like an old pair of underpants with no more give in the elastic. A whole-body scan can also show us the traces left by the ageing process on our bodies. The whole idea of 'normal ageing', entertained by many patients *and* doctors, is nonsensical. The very definition of ageing is that damage has accumulated where once there was none.

I think that for most people, 'normal' ageing means that the damage they have accumulated feels 'typical' for their age. An example: if, at the age of 45, you are trying to find your way on a map in the dark and you realise that you can't read it, it's time to get a pair of reading glasses. That's because the translucent proteins in the lens of your eyes are damaged and have lost their elasticity. Then you need a pair of glasses to correct the problem. People consider that 'normal'. We have rendered this acquired defect 'normal' because it appears at the expected age. But the phrase 'normal ageing' can be misleading. If someone needs reading glasses at 38 — is that abnormal? After all, it's only seven years earlier than expected ... But the speed at which people age varies; some people can still read without glasses when they are 52, because their lenses have retained their elasticity for longer. This is just as unexpected as needing reading glasses at the age of 38.

Ailments and diseases are also labels used by doctors and researchers to recognise individuals with specific characteristics, so they can provide treatment and care. By diagnosing a condition or disease, doctors can try to make someone better or prevent infirmity from occurring.

It should not surprise us that 'new' conditions and diseases appear all the time. There will always be a good reason for recasting an existing biological phenomenon as a new condition or disease. Sometimes, this results from a better understanding of the causal mechanism; other times, defining a condition or disease is part of a new medical strategy to delay the onset of infirmity, or prevent it from occurring.

For example, the phenomenon of weakened bones in old age has been known for a very long time. This used to be considered a symptom of the 'normal' ageing process, and therefore unworthy of particular attention. But our view of 'normal bone weakening' was completely changed when medicines were developed that could slow down the decalcification that leads bones to become brittle. Suddenly, there was a need for a precise definition of this biological phenomenon, and the medical condition of 'osteoporosis' was born. Now there is a veritable fashion for identifying the condition early and treating it as a disease.

Conversely, some conditions and illnesses disappear

from view: doctors no longer diagnose 'hysteria', for example. Others cease to exist because they are given a new name: 'dropsy' is now known as 'heart failure'. But the underlying biological mechanism that causes these physical and mental problems has, of course, remained unchanged.

If we are to find ways of preventing physical deterioration in old age, it is essential that researchers and doctors realise that ageing is not normal. They must investigate the biological mechanism of ageing, and develop treatments to prevent lasting damage. Doctors and researchers will continue to label the signs of the ageing process as new conditions and diseases. This means that they will declare us to be ill earlier and earlier, which is only justifiable and desirable if it ultimately leaves us better off.

THE DEMENTIA EPIDEMIC

We find it completely 'normal' that our thinking slows down, and we notice ourselves easily forgetting trivial things as we get older. It also takes us longer to find our bearings in an unfamiliar city, and we get lost more easily. 'What was that person's name again?' we ask our partner. We do worry about such signs: 'What's happening to me? Am I getting dementia?' But they do not really affect our ability to function properly in daily life, and so a doctor would not declare you sick or diagnose dementia on the basis of such complaints. The definition of dementia is that the functioning of the brain is so impaired that day-to-day life is disrupted. And that is not the same as feeling anxious or irritated.

Does that mean, then, that minor disturbances in brain

function are harmless and unimportant? To some extent, it does. Almost everyone has difficulty remembering things sometimes — even young people for whom a loss of brain function is not an issue. It seems we sometimes demand more of our brains than they can handle at that particular moment. There is nothing special in this. The same is true of all skills, whether it's working, cooking, or running. We always want to perform better.

However, memory problems in old age are indeed often associated with a 'substrate', a small amount of damage to the brain, which can be identified on a scan. Much scientific research is done on people with and without memory problems, in order to find out which kinds of damage appearing on such a brain scan can be related to a reduction in memory performance. To put it another way, which kind of damage contributes to memory problems, and which does not?

Damage that shows up on a scan can result from high blood pressure over a protracted period, from atherosclerosis in the carotid arteries, from the remnants of a virus infection, from a lasting inflammation of the brain tissue, from clumped proteins, or from other biological processes we are still unaware of. Knowledge of the way this damage is caused can help make it possible to intervene early in disease processes. The best thing would be to prevent the damage from occurring in the first place, but that is not likely to be possible in the near future. What we can do is slow down the pace at which damage accumulates in the brain, to defer the moment when the doctor diagnoses dementia. Not smoking, not being overweight, and taking sufficient exercise keeps your heart and blood vessels healthy for longer, and all that is good for your brain, too. Your blood pressure should also not be too high in middle age, since high blood pressure can damage the tiny blood vessels in the brain, causing a rapid increase in memory problems.

Some people are mentally still as sharp as a razor in old age, and their doctors tell them, 'Your brain scan looks like that of a youngster.' In such cases, the accumulation of damage appears to progress extremely slowly. The person in question goes home reassured, and is pleased not to have been labelled a patient. No more thought is given to it, since, after all, there is nothing 'wrong'.

This is a big mistake. People whose minds remain razor sharp in old age have a lot to tell us about ageing without accumulating damage in the brain. If researchers and doctors manage to find out their secret, that knowledge will be the first step towards helping others stay sharp of mind. It is for this very reason that we at the Leiden Longevity Study analyse families whose members show no signs of illness in old age. We hope to discover what protective factors make for their above-average healthy life expectancies. This strategy is the 'reverse' of the usual method of research that involves monitoring (older) people over an extended period to find out what causes them to get sick. We did the latter in the Leiden 85-plus study. Researchers identify the risk factors among sick people.

The brain scans of many older people, do, incidentally, display abnormalities even when they do not have dementia. Damage is there, but not so much that it inconveniences them in their everyday lives. It makes a big difference whether the person's IQ was high or low earlier in life — that is, whether they had a 'good' or 'not-so-good' brain. The reason is easily understood: if you are blessed with a good set of brains, you can handle more damage before your mental capacity starts to deteriorate and you begin to have problems functioning in day-to-day life. This idea is called the 'cognitive reserve theory'. A high IQ, or a 'good brain', is a sign that you have a lot of reserve capacity. It also explains why a solid education early in life offers 'protection' against the onset of dementia.

One view is that dementia occurs as the result of a gradual increase in various kinds of damage in the brain, caused by different biological mechanisms. This contrasts with a widely held view among researchers and the public that *the* cause of dementia is amyloids. When the proteins in the body become crumpled — that is, misfolded or clotted due to damage — they can be deposited in the tissue of the brain, and damage the surrounding brain cells. This leads to Alzheimer's disease. Some families have a genetic predisposition for producing many such amyloids, or for making proteins that are quicker to misfold. The effect is the same. These people's genetic makeup means amyloid proteins are deposited more quickly in their brains than others'. Members of such families have a higher-than-average risk of developing dementia, and more members than average are affected by early-onset dementia. Unfortunately, despite a great deal of research, we are still not able to control the deposition of amyloids in the brain. This particular disease process occurs in the majority of patients who develop dementia before the age of 70. However, they make up fewer than 10 per cent of dementia patients overall.

Unlike those with early-onset dementia, the majority of patients who develop the condition in old age have a complex disorder with several biological mechanisms at play. At the end of the twentieth century, the US-based researcher Dr David Snowdon reported his findings on the post-mortem examinations of nuns who had developed dementia in old age. Some of them showed widespread amyloid deposits in their brains, others showed only limited amyloid deposition, whereas a significant proportion of them showed no evidence of amyloids at all. All of their brains, however, displayed damage — after all, that is what causes someone to develop dementia — but that damage could be related to reduced blood supply due to impairment of the tiny blood vessels in the brain. Another possible cause of damage is a stroke, which kills off a small, or even large, part of the brain.

Examination of the brains of people who did *not* have dementia before they died confirms what a brain scan would have shown while they were still alive. Their brains also showed signs of damage, sometimes just as extensive as those of dementia patients, but their cognitive reserves were clearly greater and so they did not develop the condition. Remarkably, many people who did not have dementia *did* have amyloid deposits in their brains.

From this we can conclude that amyloid deposits are not sufficient to cause dementia in old age alone, but they are rather one of several partial causes. In old age, amyloids have limited impact, because other causes, such as damage to blood vessels in the brain, play a more important part. I myself prefer to speak of 'dementia' with my patients, rather than 'Alzheimer's disease'.

The worldwide attention afforded to Alzheimer's disease did not appear out of the blue. In the early 1900s, the

German doctor Alois Alzheimer described the symptoms of a middle-aged psychiatric patient suffering from dementia. After her death, he examined her brain, and became the first researcher to describe amyloid deposits. With this, he had discovered a plausible mechanism for the cause of dementia. At the time, Alzheimer's work did not receive much attention. Psychiatrists and anatomists disputed the importance of his new discovery, preferring to emphasise the relation between blood vessels and damage in the brain. Dementia caused by damaged blood vessels is now called 'vascular dementia'. Later, dementia disappeared entirely into the background as a recognised illness, and people who became confused in old age were declared to be 'senile'. Senility was thought to be a result of the 'normal' ageing process. In the middle of the last century, many people died of senility, and little attention was paid to their problems.

The denial of the existence of dementia is a sign of the discrimination that old people faced at the time. This view did not help in the development of preventive measures to slow or completely stop the deterioration of the brain. Led by Dr Robert Butler, American researchers in the nineteen-eighties attempted to break down this fatalistic view of dementia and its inevitability in old age. Butler was a pioneer, and this enthusiastic director of the National Institute on Aging wanted to make a statement for both young and old. 'Alzheimer's disease' was finally recognised by doctors and medical-research scientists.

The positive part of the 'war on Alzheimer's' is that it put the issue of dementia back on the map. That led to much research activity. One 'side effect' of this is that many people, even doctors and research scientists, now only tend to think of amyloids when they consider dementia. Some patients get dementia solely because amyloids have been deposited in their brain tissue. Others get dementia because the blood vessels leading to their brains have become blocked. But by far the vast majority are affected by both. Now, some researchers have begun to look into the relation between amyloids and blood flow. They have found amyloid deposits not only in brain tissue, but also in the walls of the brain's blood vessels. That can cause them to become blocked or burst, with lasting damage to the tissue of the brain as a result.

Many researchers believe that the complex pathogenesis of dementia makes the disease more difficult to combat effectively. If so many factors come together to cause the problem, how can we ever find a solution? Many are sobered by the fact that the war on amyloids has still not been won. Opinion formers and policymakers in many developed countries have predicted horror scenarios, with the numbers of dementia patients increasing drastically in the coming years. This is partly true for the Netherlands, because the post-war baby-boom generation will have reached a very old age by that time. After all, dementia is a disease that principally affects old people. But those prognoses are based on the assumption that the statistical risk of getting dementia will remain the same. However, that is very doubtful. Dutch researchers have shown that the risk of getting dementia in old age was significantly lower after the year 2000 than before. Brain scans carried out after 2000 showed far less damage due to atherosclerosis, which would appear to be a plausible explanation for the reduced risk. The epidemic of cardiovascular disease has long been on the decline, beginning with a fall in the numbers of heart attacks in middle age, and followed by a drop in the number of strokes suffered by old people. Now, bringing up the rear, we see dementia figures falling for the oldest in society.

A remarkable confirmation of this general improvement in the condition of body and mind was provided by colleagues in Denmark. They showed irrefutably that the physical and mental functions of people now in their 90s are simply better than those of nonagenarians born ten years earlier. They believe this is due, in part at least, to the fact that today's old people generally enjoyed a much better education early in life. Their brains were better nurtured.

A similar decrease in the risk of getting dementia has now also been recorded in Sweden. And, finally, a large-scale population survey in the United Kingdom has prompted researchers to report a 30 per cent drop in the risk of getting dementia over the past twenty years. So dementia is not an inevitable fate that awaits us at the end of our lives. An end to the epidemic is in sight.

FRAILTY

Equating ageing with 'collecting' partial causes of disease helps us to understand the difference between chronological and biological age. When people have accumulated a lot of lasting damage, and so have 'collected' many partial causes of disease, they become frail. They are biologically old, and that is usually also reflected in their appearance. 'He looks old for his age,' people will say. Physical appearance is also the first thing doctors look at when a new patient comes into their surgery: how old do

I judge him to be? Only then do they check the patient's date of birth to see whether their estimate was right. Sometimes patients turn out to be older than thought — this usually has something to do with the reason they are visiting the doctor — but other people sometimes look amazingly good for their age. Perhaps they have been able to avoid damage during their lifetimes, or they may be blessed with a better capacity for repair.

This initial estimation of biological age helps a doctor get a feeling for whether a patient is generally 'alright' or 'not alright'. If someone is biologically young, their chance of falling sick, becoming infirm, or dying in the near future is considerably reduced, because they have collected fewer partial causes. But judging whether someone is 'alright' or 'not alright' is difficult on the basis of appearance alone.

I recently met one of my neighbours on the street. He is in his late eighties, and has been married for more than sixty years. He and his wife live down the street, in the house they used to live in with his wife's grandmother. When the grandmother died, the couple inherited the house. They are of the pre-war generation, amiable and doughty. When you meet this couple, you might almost think such mild manners and resolve are the secret to reaching old age. When my neighbour and I saw each other, and he beckoned me over for a word, I looked at him in surprise.

'Are you using a walking stick?' I asked him. I had never seen him with a stick before.

'Yes, I get such a pain in my shoulders from using the rolling walker. So I put it to one side and started using a walking stick.' He stabbed the air triumphantly with his walking aid. 'That's a colourful tie you've got on,' he said, pointing to my bowtie.

'Thank you,' I replied. 'Anyway, you're looking well!' Which really was the case. We exchanged a few more words, and went our separate ways.

As I walked towards my house, I glanced back at my neighbour. I thought about the fact that he had given up cycling a few months before. He just wasn't up to it anymore. The doctor in me concluded that he had become frail. He had been living for a time with a heart condition that was difficult to treat: a leaky heart valve. There was a possibility he could die suddenly of cardiac arrest at night.

Frailty is a key concept in geriatric care. It's a familiar story: 'He was in such good shape. He lived alone and was able to fend for himself, hardly ever went to the doctor, but once he went into hospital, everything went downhill and he was dead within three weeks.' These are cases where 'one thing leads to another', when a broken hip or a bout of pneumonia triggers a series of events leading to eventual loss of life. Doctors and scientists define frailty in old people as the occurrence of complications — including death — following an event that would hardly cause problems for a young patient. The very word 'frail' comes from the same Latin root as 'fragile', which describes the medical condition well.

Old people are literally and figuratively frail and fragile — some more than others, but certainly more than when they were young. For example, when people over 50 go jogging, and play tennis or other sports, they render themselves susceptible to injury. If they run a little too fast or train too often, their knees, backs, or shoulders refuse to play along. What seemed to be a normal amount of stress now leads to a torn tendon or muscle, which would never have happened when they were younger. Tendons, muscles — everything has aged.

Ageing leads to a greater likelihood of a negative outcome than before. It bothers us that we do not always notice when our bodies or brains begin to deteriorate, and that we can suddenly become sick and die. So we want to know *how* frail we are. When we know that, we can take action or adapt to the situation, so that sickness and death no longer come as a surprise. Perhaps more importantly, that knowledge lets doctors know when they need to be on their guard, when a treatment is likely to have side effects, and when an operation makes no sense. In cases of cancer, for example, doctors and patients need to weigh up the benefits of a course of chemotherapy against the expected side effects. A good way of measuring frailty would help doctors and patients enormously when deciding whether an operation or a course of chemotherapy is worth it or not.

The big problem is that we do not yet have an adequate way to quantify a patient's level of frailty. We can only conclude in retrospect that someone *was* frail, and that surgery should not have been performed. Sometimes after an intervention, we consider it a 'miracle' that someone is still alive, and conclude that the patient must have had more reserves than we thought. And that we should be happy about the result.

Much scientific research is currently being done to find a way to quantify old people's level of frailty by using tests and questionnaires. Doctors hope this will help their 'clinical eye' and help them adapt their medical actions properly: continue with treatment when possible; be cautious when necessary. Dozens of tests, questionnaires, or combinations of the two already exist, and still doctors and researchers are trying to get it right. They have often thought they have finally come up with the right tool, but, in fact, none of those instruments are any better than the 'rule of four'.

This rule of thumb can be relied on to predict that doctors will estimate the correct level of frailty in two out of every four older patients they see, while the other two will be wrongly assessed. As a result, doctors sometimes do more than is necessary, and sometimes they do too little. This is how the rule works:

- A quarter of old people are found on the basis of tests and questionnaires not to be frail, and this appears to be genuinely the case. Their doctors then correctly reason that their current treatments can continue.
- Surveys of the next quarter of old people produce no abnormal results. Their doctors continue their treatments, but those patients turn out to be frail after all, as one complication follows another. This leads to over-treatment.
- The third quarter of old people are frail, according to the tests, but the result is false. In these cases, their doctors take the test results as an indication that they should be more reticent with treatment, for fear of producing complications and side effects. But if the treatment had been continued, those adverse effects would never have occurred, and medical intervention would have been avoided for no good reason. This results in under-treatment.
- Finally, the last quarter of old people will be deemed

by the measuring tools to be frail — a correct assessment. On the basis of this result, their doctors decide not to intervene, thus avoiding (fatal) side effects.

These tests and questionnaires are sometimes remarkably simplistic. 'Have you lost weight recently?', 'Do you need help washing or doing housework?', 'Are you forgetful?' When these questions are answered in the affirmative, the assumption is that this is a sign of large amounts of accumulated damage and few remaining reserves. It is striking that the predictive power of a list of five such questions is not significantly lower than that of a lengthy questionnaire with fifty detailed points. But that's not all. If a doctor knows a patient's age and sex, he already has the most important predictors in his hands, which are almost as accurate as all the other instruments together. All of us, both doctors and laypeople, believe we are better at predicting biological age than chronological age. But all too often we are proved wrong by the 'rule of four', and we still lose older patients unexpectedly.

For doctors and scientists, it is intolerable that they are so bad at estimating the biological age of patients, because this ability is so urgently needed in medical practice. For this reason, a whole new scientific tradition has arisen, aimed at identifying so-called 'biomarkers' — substances in the blood that give a better indication of the biological condition of the body than existing questionnaires. Despite many efforts, this research has so far yielded only scant results.

Other researchers take a completely different tack. They reason that frailty only comes to light when the body is exposed to stress, when the existing equilibrium is disturbed. That is correct, because when an older person is in equilibrium, little appears to be wrong with her. This has led some researchers to expose people to a 'quasi-intervention', or, to put it another way, to disturb their equilibrium and observe how they react. The predictive power of such a 'stress test' may well be higher. For example, the predictive power of a simple walking test — where subjects are asked to cover a distance of twelve metres as quickly as they can — appears to be just as high as that of all currently available questionnaires. Since this walking test has been carried out on people all over the world, after which the subjects were monitored to see whether and when they died, we now have detailed tables showing the risk of mortality as a function of age and walking speed — separately for men and women, of course.

9

THE BIOLOGY OF AGEING

The evolutionary explanation for why we age is that people invest in fertility at the expense of their own body. There are countless biological mechanisms that explain how an accumulation of damage in our body can develop into a medical condition, disease, or infirmity later in life. Often, no strict differentiation is made between the 'how?' and the 'why?' There can be no doubt that human ageing cannot be prevented. But ways of repairing the damage that causes ageing do offer new prospects for the future.

In summer 2012, the Dutch theatre company Cowboy by Night staged a performance called *Tinbergen's Gulls*. It was a tribute to Niko Tinbergen (1907–1988), a Dutch biologist who wanted to understand more about animal behaviour. The premiere performance took place on the island of Terschelling, on the edge of the nature reserve where Tinbergen had set up his field laboratory more than half a century earlier, to study the behaviour of herring gull chicks. At that time, behavioural science was a completely new branch of biology, and Tinbergen is seen as one of the founding fathers of behavioural biology. He and two of his colleagues from Austria were awarded the Nobel Prize for Physiology in 1973.

Tinbergen's experiments on the island go back to the nature-nurture question: the debate among scientists about whether particular biological phenomena can be explained by innate characteristics, or whether they are the result of environmental influences. Tinbergen was fascinated by the behaviour of young gulls. Newly hatched chicks immediately start to peck the red spot on their parents' beaks. The parents react by regurgitating food for the chicks to eat. Tinbergen theorised that the hatchlings must have a pretty good idea of what a herring gull's beak looks like before they emerge from the egg. To test his idea, he made a set of cardboard gulls' heads: some with a red spot on their beak and some without; some with different-coloured spots; and some with a spot on the forehead rather than the beak. He expected to find that chicks most often pecked at the cardboard heads that most closely resembled the real thing. And he was right. The chicks' pecking behaviour depends on the interplay between the parents, who stick out their heads, and the chicks who recognise the beak without prior experience of it. Stimulated by their parent's beak, the chicks engage in pecking behaviour, which in turn stimulates the parents to regurgitate food. Just as with all biological phenomena, it depends on a combination of nature *and* nurture.

That parent birds and chicks should be fixated on each other in this way is not difficult to understand. It increases the chicks' chance of survival, and so increases the fitness of the whole species. It also explains why this behavioural pattern has evolved over many generations of natural selection and is controlled by the birds' genes. Biologists refer to a biological phenomenon that is logical and explicable from an evolutionary point of view as an 'ultimate explanation'.

The gulls' instinctive behaviour is the result of a whole chain of events. It begins with the chick's visual recognition of the red dot. This recognition requires the stimulus to be transferred from the chick's eyes to its brain, which then must process the information. This is followed by a complex coordination of the chick's body, resulting in its pecking at the beak of its parent. The pecking triggers a gag reflex in the parents, which regurgitate food. The chick then eats the food. All these events are controlled by a complex interplay of nerves, muscles, and organs. Biologists call this a 'proximate explanation'. A characteristic of such an explanation is the description of the mechanism that explains the pecking or gagging behaviour. This requires an examination of a whole series of events, from perception of the stimulus to the execution of the behaviour. That examination provides an answer to the question of *how* it happens. The ultimate explanation answers the question of *why* the animals display this behavioural pattern.

Behavioural research is not only carried out on gulls, but also on other birds, on mice, rats, and on our close evolutionary relatives, the apes. Scientists believe there is no fundamental difference between species when it comes to the control of behaviour.

This idea was also the starting point for the theatre performance on Terschelling. During the show, the actors reconstructed Tinbergen's experiments in a hollow between the dunes. Then they appeared with bright red lips, shiny high heels, and huge, brightly coloured false hips, and proceeded to simulate sexual acts. The actors wanted to show the audience that our behaviour is also controlled by primeval reflexes and external stimuli. The performance ended with the actors covering up the erotic stimuli, and 'returning to their senses'. As if to say, 'Don't trust your instincts!'

There are also ultimate and proximate explanations for the ageing process. The ultimate explanation for why we age is that we invest in fertility at the expense of our own body. Alongside this one evolutionary explanation, there are countless types of damage that cause a reduction in the functioning of cells, tissues, and organs. Every causal mechanism of disease and infirmity in old age is a proximate explanation for how our body begins to falter. The difference between proximate and ultimate explanations is important because hundreds of theories of ageing have now been formulated. The question is what exactly each theory explains. There are popular theories about the breakdown of proteins, DNA damage, chronic inflammation, a shortage of stem cells, and the crucial role of free radicals and telomeres. The list goes on, and new mechanisms are added to it regularly. I will discuss some of the more high profile of them.

Although most of these theories are presented as *the* theory of ageing, none of them explain why we age. Thus, they are proximate explanations at best. Furthermore, closer inspection reveals that the indicated mechanism is only able to explain some of the phenomena of ageing. In brief, the biological mechanism described is nothing more than just one of the many types of damage to our body that come together to form an explanation of how we age. The sobering conclusion is that unravelling one biological mechanism is not going to allow us to counteract the human ageing process.

ACCELERATED AGEING

Babies and infants are vulnerable: they are susceptible to low and high temperatures, infections, and accidents. This susceptibility reduces considerably as they get older and continue to develop. With the passing years, their bodies gain strength, and their risk of dying from starvation, dehydration, and accidents falls. Children's chances of catching many illnesses also go down as they grow up. Resistance to infectious diseases is a good example. Anyone who has had the mumps or measles cannot get them again: once they've been fought off, children become immune to these illnesses. The reason for this is that our immune system is so highly developed that it recognises an 'enemy' the second time around, produces specific antibodies against the virus, and activates immune cells to kill off those pathogens. All of us catch a whole range of infectious diseases in childhood, which helps our immune system adapt even more closely to our environment.

Since our immune system ages as we do, the probability of getting an infectious disease rises again in old age. Babies and old people have the highest risk of dying from flu. Worldwide, seasonal epidemics are estimated to result in about 3 to 5 million cases of severe illness, and about 250,000 to 500,000 deaths. Vaccination programmes have been put in place to help vulnerable individuals muster sufficient protection against the virus. Without a flu shot, their defences would kick in too late, or not at all, with all the consequences that entails.

Some children are born with an 'immunodeficiency' — that is, an inability to produce antibodies or immune cells. From the outset, they suffer from serious infectious diseases that cannot, or can barely, be tackled with antibiotics. They do not have a normal childhood. Complications arise early in their lives, for example when a pathogen has established itself in their lungs or brains. An accumulation of damage takes place, and these children appear to age quickly. Their condition is a 'developmental disorder', and this term indicates where the solution can be found: in repairing the congenital defect that causes it. Advances in medical technology mean some immunodeficiencies can be treated effectively, for example by administering antibodies or, more experimentally, with a bone-marrow transplant. It is comparable to the way a repairperson is able to permanently solve a problem with a brand-new machine by replacing one component.

Some children appear to develop normally, until something suddenly goes wrong — the expected growth spurt caused by the sex hormones that trigger puberty does not appear. As a result, they remain considerably shorter than their peers. When they then also start to go grey or lose their hair in their twenties, and their skin begins to develop blemishes and ulcers, they appear suddenly to have turned into old people. This is compounded not long after by the fact that their voices become hoarse and weak, they lose subcutaneous fat, and they end up looking 'elderly'. When later symptoms including cataracts, adult-onset diabetes, and osteoporosis set in, it becomes clear that these individuals do not have a long life expectancy. Such patients look 80 when in fact they are only 30 or 40 years old. Most die around the age of 50 of a heart attack or cancer.

This pattern of development, illness, and infirmity is called progeria — accelerated ageing. Most progeria patients suffer from Werner syndrome, named after the German ophthalmologist Otto Werner, who described a young patient in 1906 with a type of cataract normally only seen in 80- and 90-year-olds. The symptoms of Werner syndrome resemble the normal signs of ageing, but they appear much earlier in life. A closer examination of the symptoms reveals ever more differences from usual ageing. Although the physical effects of Werner syndrome are considerable, sufferers' brains are spared. This is very different from the ageing we all experience, which damages all our organs, and which so often entails memory complaints.

Werner syndrome is an example of 'segmental ageing', in which not all of the body's cells and organs are affected by the disease process. In addition, when these patients get osteoporosis, it appears in unusual places, such as the long bones, rather than in the hip or the spinal column. Sometimes, patients with Werner syndrome develop various kinds of cancer during their lives, but it is usually of a rare type that begins in the connective tissue.

Werner syndrome is rare, affecting an estimated 1 in every 100,000 children born. It is more common in island environments, such as Japan and Sardinia. This indicates that it has a genetic basis, because parents in small communities are often related to each other; when that is the case, genetic defects are more likely to be expressed. DNA testing has identified the gene associated with this syndrome. All human cells contain 46 chromosomes, and the genetic code for 'Werner protein' is found on chromosome number 8. This protein is involved in the 'unzipping' of the DNA molecule so that the code it contains can be read by the protein-producing machinery. If that unzipping does not take place, the code remains hidden and there is nothing for the machinery to read, so the cell stops producing proteins.

In order to study the effect of damaged Werner protein more closely, scientists have taken cells from the skin of patients with the syndrome, and cultured them in the lab. These connective tissue cells behave abnormally in that they suddenly stop growing after a few dozen divisions. Researchers believe this behaviour may be the reason for an inability to replace cells sufficiently, causing the premature baldness seen in Werner patients, for example. Some cells display unregulated growth, which might explain the occurrence of cancer in the connective tissue of people with Werner syndrome.

Now that the genetic defect responsible for the syndrome is known, test models have been developed that involve deliberately inserting this mutation into the genetic material of test animals. In this way, researchers hope to unravel more details of the mechanism that causes the syndrome. There is as yet no targeted treatment programme for Werner sufferers — they depend on the possibilities offered by medical technology to treat or prevent the complications and infirmities that arise early in their lives. Replacing a lens clouded by cataracts is one such measure.

It is interesting to investigate what patients with Werner syndrome can teach us about the usual ageing we all undergo. We learn from them that an insufficient ability to read the genetic code contained in our DNA can be reason enough for the kind of damage to occur that explains some of the phenomena associated with ageing. The research in the laboratory shows that the quality of our connective-tissue cells is closely related to the development of ageing-related disease. However, this does not mean we can also conclude that the reverse is true: that the usual ageing process in humans can be ascribed to a flawed Werner gene. The number of people who carry the Werner gene defect is extremely limited, and ageing is a universal phenomenon. So we must conclude that other causes underlie the usual ageing process.

There are other, much rarer, and more extreme, forms of progeria. For example, one in several million babies is born with Hutchinson-Gilford progeria syndrome, in which the segmental ageing process takes place within a space of just ten to fifteen years.

Patients with progeria make a huge impression on us: at a (very) young age, these children look almost like 70- and 80-year-olds. But that little word 'almost' is key. Progeria is a congenital abnormality that makes these little patients sick. This is different from the usual ageing process that occurs due to damage, of various origins, which begins to accumulate gradually in the body after puberty.

OXYGEN RADICALS

One of the most popular theories on ageing has to do with the damage that free radicals can cause to the tissues in our body. Many cosmetic products are developed on the basis of this hypothesis. Radicals are atoms or molecules that have a single, unpaired electron in the outermost shell. Electrons, however, like to exist in pairs, so radicals will often bind aggressively with any biological structures in their neighbourhood. When a radical binds to another molecule, it steals an electron from it, causing the affected molecule to become a free radical itself, and a chain reaction occurs. This creates a connection between the entire internal structure of cells and tissues. It is somewhat similar to the way rust forms: iron combines chemically with oxygen, and the rusting process eats up more and more of the original iron. This is also exactly what happens in the ageing process.

One example of a typical radical is superoxide — an oxygen radical — which arises as a by-product of metabolism in cells. Hydrogen peroxide is another example; we know it as an aggressive substance used to bleach hair, for example. Radicals are used by our immune cells to kill off pathogens. At the same time, this immune reaction causes damage to our own body, which then switches on a large number of so-called 'antioxidants' to deactivate free radicals. One example of such an antioxidant is ascorbic acid, more popularly known as vitamin C.

The relation between oxygen radicals and ageing

was first reported in the nineteen-fifties by Dr Denham Harman, an American gerontologist. It was known that, although it is absolutely essential for life, oxygen at high pressure can cause side effects. This was discovered when high concentrations of oxygen were added to the air given to patients receiving artificial respiration. Their lungs were soon completely destroyed. In combination with the 'rate of living' theory, which suggests that a body can tolerate only a certain amount of damage during its lifetime, this led to the 'free-radical theory of ageing'. The central concept of this theory is that the speed of a person's metabolism has an influence on how long she will live, just as the size of a candle's flame determines how quickly it will burn down. In the ensuing years, many researchers have demonstrated connections between the number of free radicals a body produces and the processes of countless diseases such as cancer, atherosclerosis, osteoarthritis, diabetes, as well as degenerative neurological diseases, including dementia.

Radicals can cause irreparable damage to DNA, and that is the first step towards cancer. Radicals also damage the protein called elastin — the 'elastic' in our skin — making it less taut as we age. And many more kinds of damage exist. There is absolutely no doubt among scientists that oxygen radicals harm the body, but the crucial question — which has not yet been answered once and for all — is to what extent reducing oxidative damage can slow down the ageing process and prolong our life.

Although radicals are extremely unstable — they arise suddenly and disappear again just as quickly — our body has a large number of antioxidants available to catch and neutralise them before they can cause damage. Recently, experiments were carried out with nematode worms and fruit flies, in which their bodies' ability to produce antioxidants was switched off using genetic manipulation. It was to be expected that this would lead to an increase in the number of free radicals in the body, causing more damage to accumulate, and thereby shortening the animal's life. However, the last effect did not turn out to be the case. In most cases, the genetic manipulation did not have any effect, and some of the experiments even resulted in seemingly longer lifespans for the worms and flies.

There is a huge amount of interest among the general public in antioxidants that occur naturally in food, such as vitamin A, vitamin C, vitamin E, and beta-carotene. These substances form the basis for many diets and approaches presented as 'healthy'. The Moerman Diet is a well-known diet from the Netherlands, and the US-based Nobel Prize laureate Linus Pauling, and others, have advocated taking large amounts of vitamin C or E.

However, while there have been many studies that show a connection between the use of vitamin-rich food and health, they do not allow us to infer a direct causal link between the intake of extra vitamins and health. To this day, many, many randomised studies in which half the subjects are given pills containing extra vitamins, and the other half receive pills containing no vitamins, have delivered *no* consistent proof that taking vitamin supplements has any advantageous effect at all. Some studies have even shown an increase in mortality among those who took supplementary vitamins. All in all, the free-radical theory of ageing should not be consigned completely to the realm of fairy tales, but it remains the case that there is no hard proof of the theory that ageing is simply caused by a lack of sufficient antioxidants in our bodies or our food.

INSULIN AND GROWTH HORMONE

The first results of experiments with nematode worms had a massive impact when they appeared in the nineteen-nineties. Nematodes normally live an average of twenty days in the lab, but after a minuscule alteration in their DNA the worms were able to live to twice that age. It was not only their average lifespan that increased to forty days; their maximum lifespan also rose, to more than fifty days. Soon after, other laboratories confirmed these original findings. Papers were also published detailing other, similar DNA manipulations that gave rise to a longer life expectancy. It led to the identification of a number of genes that interact with each other to form a 'signalling pathway' and determine the lifespan of nematode worms. When different worms were crossed with each other, creating diverse combinations of gene variants, their offspring were able to live between four and eight times longer. Until then, no one had ever dreamt that an animal's lifespan could be influenced so strongly by just a handful of genes. This discovery led to a boom in scientific research.

The proximate explanation for this increased longevity was found in the fact that those genes regulate the worm's *dauer* stage (see Chapter 1). Nematodes can take different forms, somewhat like caterpillars and butterflies, which are also different forms of the same species. In its *dauer* form, the nematode worm is long-lived, but cannot reproduce. The ultimate explanation for the nematodes' longer lifespan is that they can enter their *dauer* stage when conditions are harsh. Not reproducing while they wait for the circumstances to change for the better is a strategy that increases each individual's chance of survival, and increases the fitness of the species as a whole.

Scientists thought at first that this mechanism was exclusive to nematodes, and that their findings could not be extrapolated to other species. They were greatly surprised — again! — when they gradually managed to unravel the mechanism at the molecular level. The signalling pathway that provides the worms with a *dauer* form was not unique, and showed a high degree of similarity with insulin and growth hormone. These are signalling substances that are used by other organisms, mammals, and humans to regulate metabolism and development.

This similarity was thought to be an indication that the signalling pathway can have similar life-prolonging effects in higher organisms, and was gradually confirmed in the years that followed. First, it was shown that a reduced level of activity of insulin–growth-hormone signalling in fruit flies leads to smaller individuals with a longer lifespan. However, the effect turned out not to be restricted to insects. The genetic experiment was repeated on mice, and they also lived longer and were usually smaller when their insulin–growth-hormone signalling level was reduced. Now it is clear that the *dauer* signalling pathway is 'evolutionarily conserved', which means that the same biological mechanism is present in several species. However, its effect varies from species to species. It is large in nematodes, less so in fruit flies, and relatively small in mice; those animals saw their lifespan prolonged by approximately 50 per cent.

The question is, of course, what effect this mechanism has in humans. Insulin and growth hormones regulate growth and development in our species. They are involved in the metabolism of sugars and fats, and control energy storage in fatty tissue. These hormones influence a whole range of processes that are crucially important for life. But do insulin and growth hormone also influence longevity in humans?

The first indication that reduced growth-hormone signalling activity can be beneficial for humans came from observations of patients with a rare congenital defect known as Laron syndrome. This defect occurs particularly in families where the parents are related. On average, one child in every four will have the condition in affected families; if both parents are carriers, a child receives a mutated gene from each of them. The mother and father do not suffer from the condition themselves, since they have one functioning copy of the gene and one mutated copy — all humans carry two copies of each chromosome, except the Y-chromosome in men. The syndrome is caused by the fact that the molecule to which growth hormone binds is damaged, and the growth-hormone signal does not get through. Hence, affected children do not develop properly, both physically and mentally, and they remain small. It might be expected that the affected family members, with their disturbed development, would die early, but many have a reasonable life expectancy. It is particularly striking that people with Laron syndrome almost never get diabetes or cancer, which are precisely the conditions that typically plague us in old age. In this respect, these patients are probably somehow similar to long-lived dwarf mice or tiny Brandt's bats, although those animals' growth-hormone signal is disturbed at a different location. All indications are that animals which can get by on fewer growth hormones live longer, but are smaller than normal.

In Leiden, the research group led by Professor Eline

Slagboom and myself investigated whether a genetically determined variation in insulin and growth hormone also affects the rate of ageing among the general population. We examined the genes connected with insulin and growth hormone in samples taken from people who had reached extreme old age. With the rapidly advancing possibilities for genetic analysis, we were able to detect subtle genetic differences among our 85-year-old subjects. These variations, which they had received from their parents thanks to the mechanism of sexual reproduction, can be seen on the biological level as the results of unintentional genetic experiments. Subjects who were carriers of one or more of these subtle variations that resulted in reduced insulin and growth-hormone signalling were smaller and had a lower mortality rate, just like worms, flies, and mice. Other researchers observed similar results in their studies of centenarian subjects.

This shows that the evolutionarily conserved mechanism that affects the lifespan of worms, flies, and mice is also active in humans. However, unlike genetic experiments carried out in the lab, natural genetic variation within the general population is small, and its effects are much more limited than they are in worms and flies.

In Leiden, where we have collected data on a large number of long-lived families, we have investigated whether those people were carriers of a genetic variation that results in less active insulin–growth-hormone signalling. We were not able to establish a direct link. This result did not come as a great surprise. In the course of our research, we had already established that the offspring of these long-lived families look exactly the same as their partners, whom we considered 'normal'. They were just as tall and just as fat — that is, they were not smaller, as was the case with the long-lived worms, flies, and mice. But how does this negative finding among long-lived families relate to the positive findings among 85-year-olds? We think that the long-lived families are endowed with another — genetically determined — biological mechanism that enables them to lead longer and healthier lives than the average. There are apparently various signalling pathways — different proximate explanations — for the same beneficial effect on the condition of our bodies. Sometimes, it seems these signalling pathways do overlap and interlock. For example, we were able to show that the offspring of long-lived families are less likely to develop diabetes, and have lower blood-sugar levels and a better metabolism. Those are precisely the biochemical characteristics shown by long-lived mice in the lab.

It is noteworthy that less active insulin and growth-hormone signalling is generally associated with beneficial biological effects and a longer lifespan. Insulin and growth hormones are absolutely necessary for normal development and survival. However, there are clearly negative aspects to insulin and growth hormones when it comes to old age. This is another example of 'antagonistic pleiotropy' (see Chapter 6). Inflammation is a further example of this phenomenon, since an active immune system is good for survival, but can lead to undesirable side effects in old age. But we should not be surprised that our bodies are so badly adjusted to old age. After all, it is only the beginning of our life, the periods of development and adulthood, that have been optimised over countless generations of natural selection. In evolutionary terms, the end of our lives is something of an afterthought.

What we definitely should *not* do is to doggedly insist that everything should remain the same as it was in our young adulthood. We know that enhancing a low level of growth hormone in old age to raise it to the level at puberty does not lead to improvement. The muscle growth this promotes may result in a stronger-looking body, but there is no corresponding increase in muscle strength, and taking growth hormone throws the body's sugar metabolism out of kilter. It may also very well be the case that using growth hormone increases the risk of cancer. A similar effect has already been proven conclusively for hormone-replacement therapy, when menopausal women's oestrogen levels are artificially boosted to 'normal' levels.

Our bodies and brains are complex systems that will undoubtedly be able to be more fine-tuned in the near future — especially in old age. But little good can be expected from the supposed panaceas available now. A person does not remain healthy for longer by taking hormones, vitamins, amino acids, or minerals. They should only be taken when a serious deficiency of one of those substances is detected in the body.

SHOULD WE EAT LESS?

One theory of ageing that is rapidly gaining in popularity is that we can stay alive longer by eating less. While we used to have to face regular food shortages in our evolutionary past, advances in agriculture and animal farming in developed countries mean we now live with an abundance of food that is forced on us day and night. Shortened lives due to food scarcity have been replaced by disease and death from overeating. We have to try to avoid overeating and the resultant obesity by resisting the many temptations that surround us, by consuming fewer calories and burning more. We now burn far fewer calories than we used to, because we engage in far less physical exertion. In other words, our modern environment no longer fits with the set of biological tools we evolved with. There is no need to be emaciated, and particularly not in old age, but young and middle-aged people have the lowest risk of illness and death if they are in good physical shape, while older people have the best chance of survival if they are slightly chubby. It seems advantageous to have some (fat) reserves in old age, in case of an internal or external 'attack' in the form of illness or accidents. No one has (yet) come up with a better explanation for this phenomenon.

True proponents of eating less, incidentally, do not advocate chubbiness in later life; they dream of living longer by eating 20 to 30 per cent less than the recommended average calorie intake. This results in being underweight, constant hunger, and a heightened sensitivity to cold. People who follow this strict regimen claim it gives them a feeling of wellbeing.

The whole idea behind this so-called caloric restriction is based on laboratory experiments conducted on mice, which live longer when fed 30 per cent fewer calories. They demonstrably stay healthy for longer and die later than mice that are allowed to eat as much as they want. This effect was first described in the nineteen-thirties, and is one of the most studied mechanisms to counteract ageing.

Caloric restriction works not only for mice, but also for other test animals, including fruit flies. Although the beneficial effect has been demonstrated repeatedly and is an expression of a genuine biological mechanism, it does not work for all flies and all mice. For a large number of genetic variants, caloric restriction has no consequence, or may even have a life-shortening effect. This points towards a 'nature-nurture' phenomenon, since the effect of caloric restriction (nurture) depends on the test animals' genetic traits (nature). It seems each genetic variant within one species has its own optimum calorie intake. If that is the case, too much food is bad, but so is too little. This has now been proven in experiments.

Of course, what we really want to know is how this applies to humans. Is it better to eat (far) less food than the recommended amount? The answer may be found in a long-term experiment with rhesus monkeys, a species that is closely related to us in evolutionary terms. In these experiments, the monkeys are randomly separated into two equally sized groups, one of which receives a normal amount of food, while the other is given only 70 per cent of that amount. (Only the number of 'empty' calories is restricted.) The food is all of the same quality, sometimes with added supplements to make sure the test monkeys do not suffer from deficiencies of vitamins and other essential nutrients. The experiment is being carried out by two separate institutions in the United States, each working independently. It has been going on for several decades, and the effects on life expectancy can now begin to be measured, because a considerable number of monkeys have now died, and the general risk of mortality can be estimated for each group.

So what are the results so far? The monkeys that have eaten calorie-restricted food over a long period of time look younger in appearance, have more energy, and are less likely to develop the diseases that are typical of ageing, such as diabetes. This suggests that the ageing process is slower in those animals. However, this does not (yet) appear to be reflected in their survival rates. When all causes of death are taken together, the monkeys on the caloric restriction regimen do not live demonstrably longer. The studies are still underway, and the final word has not yet been spoken. These unparalleled experiments will need more time to show their full potential.

The quest for the Holy Grail is not only a matter for mythical kings like Arthur. Many research scientists hope to find a magic elixir that will stop the ageing process. That is an unrealistic quest, however. The number of biological mechanisms that cause damage to accumulate — proximate explanations — is endless. This means that there is no single way to prevent ageing. Some researchers are now advocating a change of course, with more focus on finding new methods of repair and replacement — that is, more external intervention when the body fails from within. Such methods have already come a long way. We hardly bat an eyelid these days when someone gets a replacement lens in their eye, or a new hip. There is hope that medical technology will soon provide a solution to childhood diabetes, which is caused by a loss of the insulin-producing cells. The islets of Langerhans, the source of those cells, can now be replaced. As this shows, we are moving ahead, step by step.

10

LONG MAY WE LIVE

It is many people's wish to remain in good health while they grow old, and then to die suddenly. But that is seldom the way it happens. Human life often has a 'ragged' end: a period of increasing infirmity eventually leading to death. Preventive measures and early intervention means we are sick for a greater net period of time, but infirmity and death come later. Thus the years of increasing impairment have not disappeared; they have been postponed until later in life. The Dutch prime minister responsible for introducing state pensions, Willem Drees, said, when the law was introduced, that the pensionable age should always be pegged to life expectancy. If we were to follow his advice, men should now continue working until they are 70, and women until they are 72. Comparable figures would be appropriate in the rest of the developed world.

The average Dutch woman spends half her life sick. Shortly after her 40th birthday, she develops her first chronic illness, with an average of 43 more years to live. Illness among women now appears a good ten years earlier than it did 25 years ago. The crossover point to 'half a life of illness' was reached just before 2010. The striking thing is that Dutch women in their forties do not feel sick at all, nor do those in their fifties and sixties. Two-thirds of 55-to-85-year-olds describe their state of health as 'good' or 'very good'. Fewer than one in ten describe their health as 'bad' or 'very bad', even at the age of 85. It is only as they approach the age of 70 that around half admit that they can no longer see, hear, or move as easily as they once could. And most feel this causes them only minor problems. The situation is similar for men.

MORE YEARS OF ILLNESS

There are some doctors who often label people 'sick' although those people do not actually feel there is anything wrong. Doctors do this in the hope of improving their patients' lives: early intervention can keep people healthy for longer. Sometimes they treat high blood pressure; sometimes they lower a patient's high cholesterol level with tablets. Patients who are having trouble remembering words, for example, are sent for a body scan. If that shows that blood clots from the heart have made their way to the brain, doctors will begin treatment with anticoagulants as a precautionary measure to prevent a stroke. Sometimes a mammogram during routine screening reveals a precancerous abnormality, and the malignant cells are surgically removed. In none of these cases did the patient feel sick, but when preventive measures and screening indicated a future health problem, they were declared sick nonetheless.

The gap between 'being sick' and 'feeling impaired' has increased rapidly over the decades. People in general, but politicians and policymakers in particular, have not really got wind of the fact that doctors and researchers are now increasingly using the term 'sickness' among themselves to refer to biomedical phenomena they are trying to make progress in treating (see Chapter 8). The term 'sickness' used to have a very different connotation. A bypass operation, or the insertion of a stent — a little mesh tube — can improve blood flow through the coronary arteries in comparison to before the intervention, when all sorts of things were wrong that the 'patient' was unaware of. From a medical and biological point of view, diagnosing an illness and carrying out a repair to remove it has improved the 'patient's' life. But some people feel worse after such an intervention, precisely because they have been labelled 'sick'; they withdraw into the role of a sick person, and scale back their day-to-day lives. The situation is often made worse when their doctor tells them that they not only have a problem with their coronary arteries, but they also have high blood pressure and cholesterol levels. 'Take it easy, don't overdo it,' their family and friends advise them in hushed tones. Such advice is well meant, but has the opposite effect to that intended. Physical activity and sport bring down blood pressure, weight, and cholesterol levels, as well as improving the function of the heart, even in people with heart conditions.

When politicians and policymakers look at the statistics, they see that 50 per cent of the working population have developed a chronic illness by the age of 45. In the West, 65 per cent of the population have two or more chronic conditions by the age of 65. Many people think that raising the age of retirement is impossible in such a scenario. But just because people are labelled 'sick' does not mean they are automatically disabled or unable to work. In the past, people often did not even know when they had come down with a medical condition. Some people find that a comforting thing — 'ignorance is bliss' — but they are simply sticking their head in the sand. Early diagnosis can prevent problems in the future.

In the nineteen-eighties, during my training as a consultant in internal medicine, I was taught to prescribe lots of rest for heart-attack patients, primarily to prevent fatal disturbances in the patient's heart rhythm. But this policy was also aimed at making it easier to handle the extreme damage to the heart tissue that is caused by a myocardial infarction. Many weeks in hospital were followed by a rehabilitation programme taken at a very cautious pace. Often, there was no question of a complete return to work, or even to the patient's previously normal life.

At that time, you were labelled sick, and you felt sick. It rarely comes to that nowadays because most imminent heart attacks are nipped in the bud. Intervention comes immediately after the first symptoms appear, blood flow is restored in the coronary arteries, and fatal tissue-damage is averted. The causal chain of events is broken, with amazing results. Most heart patients who develop acute problems are usually sent home after just a few days in hospital. They return to normal life as quickly as possible, and are encouraged to live more healthily, under professional guidance. Their symptoms have disappeared, and the problem has, for the most part, been solved. But, still, some people who have been stuck with the label 'heart patient' often ask themselves despairingly whether they should now be feeling sick or not.

Activation, rehabilitation, and normalisation are the magic tools of modern medicine. Midwives were the first to realise how to prevent the problems that can occur before, during, and after birth. Childbirth used to be a dangerous matter for young mothers. Formerly life-threatening complications such as bleeding, infection, and thrombosis were common.

'See this book, ladies and gentlemen?' said my obstetrics professor to us students during a lecture, holding up the standard work on complications in childbirth. 'Almost all the dangers it contains have disappeared because we now get young mothers out of bed from day one, instead of prescribing a week's bed rest.'

Since then, that lesson has been introduced in many branches of medical practice. Stroke patients now begin rehabilitation within 48 hours, rather than waiting until everyone has recovered from the shock. This rapid-reaction approach can improve the patient's residual functional capacity. Old men or women who take a fall and have their hip surgically replaced with an artificial one made of steel should preferably be back on their feet within 24 hours. This prevents the onset of pneumonia and thrombosis — complications that often used to be fatal. Older hospital patients who become confused due to an acute condition benefit greatly from being encouraged to get out of bed and to dress themselves, and by being surrounded by familiar faces. As the saying goes, 'Use it or lose it'.

What is true on the hospital ward is equally true in the outside world. Many people are surprised to learn that experimental evidence suggests that regular physical exertion is beneficial, even in cases of heart failure. Patients who are prescribed regular moderate exercise do better than those who are not. Experimental studies on humans have now also shown that exercise has demonstrable benefits for patients whose knees are damaged by osteoarthritis, and whose mobility is limited by the pain it causes. Exercise does not reduce the amount of stress on the joint, but the exertion strengthens the mechanisms of repair and compensation, and the ultimate balance is positive. Quite apart from the benefits it brings for the heart and musculo-skeletal system, physical exercise and sport improve our mood. There is increasing scientific evidence for the proposition that physical exercise also helps prevent a deterioration in the functioning of the brain. And that is different from doing crossword puzzles and 'train your brain' exercises.

Prevention and early intervention make for longer and healthier lives, but they also cost money, and it is difficult to calculate how much is saved by preventive measures. Obviously, there is always a personal benefit for the patient, but preventive medicine can only result in material benefits for society if patients make good use of those additional years of health by remaining part of the workforce for longer, or by contributing to society in an unpaid capacity. That saves on pension payments, and brings in taxes and social-security contributions. Public spending can also be lowered by helping older people function independently for longer.

The average person, however, is likely to respond by saying 'I have a right to my pension!' And most politicians believe old people are slow and often sick. National pensions were finally introduced in the Netherlands in 1957. The man responsible for drawing up the necessary legislation, prime minister Willem Drees, included in his draft law a clause pegging the pensionable age to residual life expectancy. At that time, 65-year-old Dutch men had a further life expectancy of thirteen years; this has now risen to eighteen years. So if we were to follow Drees' advice, men in 2015 should keep working for those extra five years they've gained, until the age of 70. In 1957, 65-year-old women in the Netherlands could expect to live another fifteen years. That has now risen to twenty-two. This means, according to Drees, that women should now remain in the workplace a further seven years, until they reach the age of 72. Only if the pensionable age is linked to life expectancy, as Drees suggested, can a national pension system remain affordable.

MORE YEARS WITHOUT IMPAIRMENT

The Netherlands is one of the few countries whose national statistics office (Statistics Netherlands) has begun to draw a distinction between diagnosed illness and perceived health, between being sick and feeling impaired. The bureau describes those who have no trouble with their hearing, eyesight, or mobility as having 'no physical impairments'. Conversely, respondents in Statistics Netherlands' health surveys who report having some or a lot of difficulty with one of those three abilities, or indeed all three, are described as living with physical impairment. Statistics Netherlands introduced such surveys in 1983, which leaves us with an unique opportunity to interpret temporal patterns on a national level. There are also ample surveys of selected individuals in other developed countries that show similar patterns over time..

In the past thirty years, the average time a newborn baby can expect to live without disease has *de*creased in the Netherlands. When confronted with this statistic, many people wonder if and why the situation has deteriorated.

However, at the same time, the life expectancy without impairment has increased for newborns. The 'ragged' end of life, an average of ten years of living with impairment in day-to-day life — two years less for men, two years more for women — has stayed the same length. The years of impairment have therefore neither disappeared, as many tacitly hope, nor have they increased, as feared by practically everyone. They have been postponed, and the explanation for this is a biomedical one. By diagnosing diseases and disorders earlier, doctors make use of prevention and repair techniques to postpone the years we have to live with lasting damage to our bodies and brains. We therefore develop impairments in our day-to-day life at a greater age, and our expected life in subjective good health increases in length. This is the situation when respondents in Statistics Netherlands health surveys describe their own general state of health as 'good' or 'very good'.

It is noteworthy that the gap between diagnosed illness and actual impairment is larger among women than men, and that the number of years of sickness has risen more rapidly for women. The average Dutch man and his female counterpart both begin to report feeling physical impairment in their daily lives from around the age of 70. But women receive their first diagnosis of 'chronic illness' at an average age of 40 — eight years earlier than men in the Netherlands. This is because women develop far more musculo-skeletal complaints than men do: osteoporosis and bone fractures, rheumatism and worn joints, muscle weakness, and similar conditions. Furthermore, research has shown that women go to the doctor more often with their complaints than men. So it is hardly surprising that they also receive their first diagnosis earlier in life than their male peers. To put it bluntly, men stick their heads in the sand, and women complain more. Some researchers believe that this increased readiness to express their complaints and to visit the doctor earlier are responsible for the fact that women live longer on average than men.

Life expectancy without impairment is not only rising for newborns, but also for today's old people. When they reach the age of 75, men and women have already outlived a third of their generational peers who will have died of disease, and the prospects for such survivors are relatively good. Many people are still completely healthy at that age. In 1985, half of 75-year-old women could still look forward to four years without chronic illness, and the situation was more or less the same for men. Over the past twenty-five years, residual life expectancy without illness has fallen by two years. Once again, it is not correct to think that life in old age has gotten worse. We can easily see that this assumption is erroneous when we look at the fact that at age 75, residual life expectancy without impairment in daily life has risen from four to six years. The remaining six years with impairment — two years less for women, and two years more for men — has remained the same.

The tendency among doctors for rapid diagnoses and treatment, and the desire of patients to receive a diagnosis and treatment, deserve mention here. Of course it is preferable for disease to be diagnosed early, and for intervention to take place as soon as possible to prevent the situation becoming more serious. But, at the same time, there is a great danger of over-diagnosis and over-treatment, especially in the case of older people. This can have all sorts of negative consequences. Early identification and diagnosis can unsettle people and make them anxious. If, for example, a whole-body scan leads to the removal of a sample from a damaged area, complications such as haemorrhaging can result. Another danger is kidney failure caused by the contrast medium administered for the scan. It can be even more serious if a damaged spot is removed by a surgical intervention, but later turns out not to be malignant, or when treatment is given but does not work. In such cases, the only result is damage for no good reason. Doctors need to adopt a more restrained approach, and be less keen to make diagnoses and prescribe medication. They should take the time to explain to patients and their families that early detection is often pointless because, more often than they would like, there is no available or effective treatment.

This is the reason that there are no official screening programmes in the Netherlands for cervical and breast cancer for women over 60 and over 75, respectively. This can seem like age discrimination, and some civil groups have launched campaigns to claim their rights. However, their claim is not justified, because for older women, the benefits — the additional years of healthy life that can be expected — do not outweigh the costs of the screening. Those costs are not only of a financial nature, but are also of precisely the kind that we recognise as the undesired side effects of diagnosis and treatment. Many people suffer from these side effects, which outweigh the expected benefit to a given individual, in particular because it is impossible to predict which woman will benefit and which will not. For these reasons, most experts are not in favour of cancer screening for people over the age of 75.

A broad-based group of experts recently assessed 26 screening targets recommended for older people in the Netherlands, the United States, Britain, and Australia. For most diseases screened for, there was no effective treatment available. In such cases, screening makes absolutely no sense at all, and can even be considered unethical. Despite this, older people are increasingly subjected to testing. Tests for dementia, for example, can be a huge hazard. These tests are also used to diagnose the early stages of dementia, such as 'mild cognitive impairment'. People who receive such a diagnosis believe they are doomed to die with dementia, but that is far from always being the case. After delivering the diagnosis, modern medicine has nothing to offer them by way of slowing down any cognitive degeneration. Such memory testing is only of value if it can be followed up by effective treatment, and that is still lacking at the moment, although this may change in the future thanks to continuing developments in medicine.

The panel of experts found that it is only useful for those aged 60 and above to check whether they take sufficient physical exercise. The panel deemed it sensible to check for smoking and the risk of heart disease among 60-to-74-year-olds, irrespective of whether they are active or frail, but not for those above that age. The experts did not consider it appropriate to screen actively for hearing and eyesight problems, skin cancer, dementia, depression, anxiety or loneliness, malnutrition, incontinence, alcohol problems, chronic pain, kidney disease and diabetes, or sleeping problems. They reasoned that people can seek help themselves for many conditions if they are experiencing problems. Most people in the Netherlands have a regular general practitioner, and those aged 75 and above have contact with their doctor's surgery an average of sixteen times a year.

All those visits to the doctor are a great worry for the health system's accountants. They believe healthcare costs are skyrocketing because 'people keep getting older'. That is a misconception. It is not life in old age that is expensive, but its ragged end. There is almost always a final condition or complication that doctors will do all they can to try to reverse, although their efforts eventually fail.

The post-war baby boom is not particularly responsible for the rise in healthcare costs. According to the calculations of health-sector economists, less than 1 per cent of the annual 4 per cent rise in the healthcare budget can be ascribed to the cost of the ageing population. Rather, pay rises are responsible for a large proportion of the budgetary increase, along with the fact that doctors are now quicker to declare us sick and prescribe treatment, which is enabling us to live longer. Such added investments only benefit us as a society if we convert those extra years of health into the production of tangible and intangible assets.

THE RAGGED END

In developed countries, impairments to daily life have been postponed to old age, in parallel with a rise in overall life expectancy. Between the beginning of life impairment and the moment of death, there is a period that I call 'the ragged end', which lasts an average of about ten years. This is the period in which people experience problems with a body that no longer does what it should, and then they die.

The question is: How will this develop in the future? Will the ragged end last longer, as doctors try to keep us alive at all costs? Or will it get shorter, as doctors are better able to prevent disease, impairment, and disability, allowing us to stay healthy longer, and then die suddenly?

One hundred years ago, the situation was very different. At that time, illness and the complications it entails were barely linked to age. Anyone could be struck by misfortune. Now, the first 40 to 50 years of our lives have essentially been 'swept clean'; the majority of the time we spend being ill has been shifted to the end of our lives. The average age at death has risen considerably to around 85 years; almost everyone dies within a twenty-year span either side of that age.

In 1980, the American doctor James Fries predicted that people's ages at death would be much less widely spread in the future, and that the ragged end would be reduced to a minimum, since doctors would be able to prevent disease. We would then die at a genetically pre-programmed age. This rosy view is still embraced by many today, under the banner of 'healthy ageing'. But actual observations of people in the intervening time are not consistent with this prediction. Disease has not been prevented; disease is a manifestation of the ageing process. The lasting damage, the complications, the impairments in daily life, have, however, been postponed. The length of the ragged end to our lives has remained constant for a long time, but has been pushed back until later, along with the average age at death, at a rate of two to three years per decade.

The idea of a genetically determined lifespan has no scientific basis. There is no evolutionary programme for ageing; there is no programme for death. In line with this, we see the average life expectancy and the maximum lifespan increasing all the time. It is the random damage to our bodies that determines how quickly we age — the interplay of our genetic makeup and the environment we live in. The great variation in the speed at which different people age determines when a person dies. And there is still quite a difference between different people. Of course, we have to choose our father and mother carefully to get the right genes. And we can naturally influence it somewhat ourselves — for example, by looking after our bodies. But the rest of it is luck.

Some wonderful experiments have been carried out on nematode worms that, in a laboratory setting, are all genetically identical, like monozygotic twins. Furthermore, they are all raised in the lab under strictly standardised conditions. It might be expected, then, under such consistent circumstances, that the worms would all die within a short space of time — all else being equal. But that is not the case. Experiments showed that there was a considerable range in the worms' length of survival. The length of time between the death of the first worm and the death of the last one was equal to about a third to a half of the overall average lifespan. This clearly indicates that there is not a genetically determined mechanism for the curtailment of life at a certain age. In a subsequent experiment, the range of difference in the age at death was compared to that among a genetic strain that lived twice as long on average. The variation in age at death remained the same.

These experiments with nematodes show that it is highly unlikely that the moment of death can be reduced to just one point in life. The results confirm that damage accumulates gradually until it eventually leads to death. Random damage (bingo!) plays a large role; we have little influence over the exact moment when we die. With worms, the speed at which lasting damage is acquired can be reduced 'easily', and the average age at which the damage manifests itself can be postponed until later in life. And that is exactly what we observe in humans, too.

11

THE QUALITY OF OUR EXISTENCE

Active and healthy ageing is the new creed in the developed world. But how can we achieve that, and how can we shape our new, longer lives? If we want to know what exactly 'healthy' means, we must look beyond the outdated definition from the WHO. Old people's own perception of their state of health and the quality of their lives is extremely important. Research has shown that social contacts are crucial for their wellbeing, as well as an ability to adjust constantly to changed circumstances, including the decline in their physical and cognitive abilities. Older people who manage to do that are successful in their old age. In fact, people in old age give their lives a rating similar to the one that young people do. This means that most older people are able to deal with setbacks, illness, and impairment.

In contrast to just a few years ago, when the ageing society was touted as a great problem waiting to happen, the political reports, conferences, and research programmes of today have titles such as 'Long May We Live', 'Growing Old Successfully', and 'Healthy Ageing'. The 'problem' of the ageing population has not been solved, but these new headings suggest determination and vision, while politicians, administrators, and policymakers continue to try to hold back the tide like King Canute.

What is going on? Now that almost everyone in developed countries will enter retirement (at some time), and the life expectancy for those of pensionable age and above is increasing rapidly, the old-age dependency ratio is rising as inexorably as the tide. Society can no longer ignore the fact that more and more people are living longer and longer. Our society needs to be restructured, but the necessary changes are routinely postponed. Until recently, such a restructuring of society was not really thought necessary because we could still easily afford to pay for so many people's retirement. The new economic reality since 2008 has completely changed that, however, and has made lasting change urgently necessary. The increasing number of older people, and the increased need for care this entails, require a completely new way of using public and private funds.

To lend strength to this 'new way of thinking', opinion formers no longer talk about 'problems', but prefer to refer to 'challenges' and 'opportunities', avoiding any negative labels for old people or old age whenever possible. The European Commission advocates strongly for a serious consideration of demographic change, and, in line with this new image, the new buzzword is 'healthy ageing'.

Almost every country in the world can expect to see an increase in the average age of its inhabitants as life expectancy rises. But average age is also highly dependent on the structure of a country's population: the distribution between younger and older people. And the distribution is in turn dependent on (the fall in) the number of births and the balance between the number of people who migrate into and out of the country. Of all the continents, Europe has the highest average age, and of the major countries, Japan, Germany, and Italy top the list.

By no means are all citizens convinced that the ageing society offers 'only opportunities'. More and more people are increasingly irked by the positive tone of this politically charged debate. They experience with their own bodies and minds that the ageing process certainly does take its toll, or they see it in their partners and friends. They feel anxious, or even depressed, at the thought of having to live to extreme old age. The simple reason is that negative reports about ageing eclipse the positive sides to getting older. Our parents and grandparents accomplished a formidable feat over the past one hundred years in learning how to live healthily for longer and longer. Our challenge is to learn how we can give meaningful form to that longer life. This is why we should speak much more often about how we can turn life in old age into something beautiful. And why should we not lend an ear to older people themselves, to use their experience as a source of inspiration?

WHAT IS HEALTHY?

What opinion leaders and policymakers mean when they speak of 'healthy ageing' is not the same as what doctors and medical researchers mean by the concept. And ordinary people have yet another idea again about what it means to grow old healthily. In short, there are many ways of defining the word 'health'. One interpretation is that it is a benchmark for the state of a person's body and mind. According to this, 'healthy ageing' is the best possible state of health attainable at a given age. In 1948, the WHO defined health as a state of complete physical, mental, and social wellbeing, and not merely the absence of disease or infirmity. The beauty of this broad definition is that it indicates that there is more to health than not being sick or impaired. This was a revolutionary idea in the period immediately following the Second World War. You are only healthy if you feel good about yourself, and if the ageing process has not yet caused impairment to set in. Broadly, this describes those people who look good for their age. They are the elite few who have inherited a great capacity for repair from their parents, or who have lived healthy lives, or who have simply been lucky. They are the ones we think of when we hear the popular expression, 'Everyone wants to grow older, but nobody wants to be old.' This kind of success is reserved for the very few.

Advocates of the 'healthy ageing' doctrine take a very literal view of the pursuit of health. Since this means preventing sickness and impairment into very old age, 'healthy ageing' is a futile and depressing mission for the average older person. After all, the vast majority of them are affected by one or more chronic illnesses, making it impossible to fulfil the demands of healthiness. What at first appears to be a nice, positive phase — 'healthy ageing' — turns out, on closer consideration, to be nonsensical. 'Healthy ageing' is an impossibility, and the phrase gives people the wrong idea. We can, of course, intervene in the ageing process to slow it down, but that does not prevent damage. It postpones it to a later time in life, enabling us to remain healthy for longer.

Many people who are old don't feel it, and that is the second interpretation of 'healthy ageing'. When you ask them about being old, some react with shock and horror. They certainly wouldn't describe themselves as old or unhealthy, even in advanced old age, and are quick to point to the state of their neighbour, for example: 'Go and ask her. *She*'s old. She can't walk anymore!' Adaptation to the ageing process is central to this second perspective. As we saw in Chapter 10, there can be a great divergence between illness, impairment, and subjective health. Two-thirds of older people in the Netherlands rate their own health as 'good to very good', even the oldest of them. Fewer than 10 per cent feel their own health is 'poor or very poor'. This is remarkable, since most older people are registered at their doctor's practices as suffering from various ailments and impairments. So subjective health may be a more important criterion than those in the WHO definition.

For outsiders, including doctors, sickness means something other than it does to the sick or disabled themselves. Their subjective health is much closer to a state of 'wellbeing' or 'quality of life' than healthy people might predict. This is what researchers call the 'disability paradox': people feel good, although they are sick or disabled, according to their doctor. Many (young) people believe that their life would be over if they had to spend the rest of it in a wheelchair. But that is not the case. Every year, many people are confined to a wheelchair by accidents and illness. This is followed by a period of anger, despair, denial, and depression, but eventually almost all wheelchair users manage to build a new life for themselves. Very few take their own lives. Human beings have a great capacity for adapting to new circumstances, even when their bodies or minds let them down.

The fact that most people with an illness or impairment do not feel sick or disabled, coupled with the fact that they are able to accommodate themselves to a new situation, means that a completely different way exists of looking at the ageing process. According to the second perspective, it is possible to maintain 'healthy ageing' by continuously adapting to new circumstances, including physical and cognitive deterioration, and so avoid the subjective feeling of being old. People should make up their own minds, rather than paying attention to the 'benchmark' that people around them — doctors and medical researchers — may want to judge them by. Rather than going into denial about loss of functionality, this individual attitude is based on possessing the resilience, motivation, and energy to compensate for a deteriorating body and mind. The ageing process then ceases to be a many-headed monster that must be defeated at all costs. Rather, it is a biological fact that we can adapt and adjust to.

THE LEIDEN 85-PLUS STUDY

Adjustment to old age is studied by many researchers who are interested in older people. The Gerontology and Geriatrics Department at the Leiden University Medical Centre invited every inhabitant of the city of Leiden on their 85th birthday to participate in the research project, and from 1997 to 2013 monitored those who agreed.

Anyone could take part; no selection was made on the basis of health, functionality, or living situation. In order to better understand why older people continue to experience a high quality of life despite the illness and impairments they develop in old age, the participants were assessed according to the WHO definition of health. All the participants were visited at their homes in order to examine their level of physical, mental, and social functionality, and to assess their self-perceived level of wellbeing. This was accompanied by in-depth interviews: open, unstructured conversations to take stock of their ideas and motivations, and the factors contributing to their failure or success, as seen by the older people themselves. Data was collected in this way from 599 participants. Twice as many women as men took part, because more than half of the participants had already lost their spouse. More than 80 per cent of the 85-year-olds lived in their own home; the rest lived in institutions or assisted housing. Only a minority had studied at university level, which is typical of the time when this generation grew up. In a nutshell, these 85-year-olds in Leiden were a cross-section of the very oldest portion of the Dutch population.

Poor physical functionality was reported by 20 per cent of the participants; they were dependent on others for help with one or more of their daily activities — mainly getting dressed, undressed, and washed. Most of the older people reported only minor problems in daily life: they lived independently at home, but required help with housework, for example.

Most participants considered health to be the retention of a number of basic abilities, such as sight, hearing, and mobility, and the absence of life-threatening illnesses, such as cancer. Adjusting to the physical changes they experienced in old age, for example by taking things more slowly, was seen as natural and obvious. Acceptance of functional impairment was the way to stay feeling positive. Those who had not yet developed illness or impairment considered themselves lucky, and did not see it as a personal success. Many undertook measures to influence their functionality in daily life positively. As a precaution, they sought housing close to their children, displayed risk-avoidance behaviours (such as giving up cycling for fear of breaking a hip), or attempted to keep in good physical shape by doing gymnastic exercises or using an exercise bike. They saw an optimum physical functionality as a hypothetical ideal, since chronic illness and impairments were accepted as common at their age.

One of the participants who lived independently put it this way: 'I may be disabled, but I feel healthy.' He had been in a wheelchair since suffering a stroke 24 years earlier. His entire house was full of practical aids. It took him the whole morning to get up and make his bed, but he was proud of his ability to do this without outside help. He, too, showed that acceptance and adaptation were essential for his feeling of wellbeing: 'Yes, I'm dependent on others for some things in my life, and nobody can change that, but I intend to try to stay independent in other aspects of my life for as long as I can.'

All subjects were given a memory test. Around 20 per cent showed probable signs of dementia. Some already had such a diagnosis; others had never been examined for dementia. Three out of every five participants reported no serious problems with cognitive functions. Most participants were afraid of memory loss — 'Have I got dementia?' — because they feared losing their personality. Good mental functions were held in high regard, and some subjects invested in maintaining them by doing memory-training exercises. Some felt sad and dejected at the loss of their partner. An additional 20 per cent of the participants reported in the questionnaire that they suffered from feelings of depression. This was clearly connected to problems with social interaction. When asked about loneliness, 16 per cent reported feeling isolated.

One particular participant reported feeling 'muzzled'. The woman in question was a widow, and had outlived four of her six children. Whenever she was unable to sleep at night, she would talk to their portraits in the living room. She was restricted in her mobility, was in constant pain, and spent most of the time at home in her apartment. She was sad, not just because of the pain and the loss of her loved ones, but more because she had no opportunity to tell others what was going on inside her. She craved support and sympathy, but nobody wanted to listen to her complaints. Her son had told her, 'Mother, stop telling sad stories all the time. I want you to be happy!'

All subjects were asked about their social lives. Roughly a third took part in one or two social activities per week, most commonly in the form of receiving visitors. At the other end of the spectrum, a third of the old people were active in various social networks. They had visitors very often or went out visiting themselves, spoke on the telephone with friends, played group games, or participated in clubs and church services. For most subjects, social functioning was crucial to their sense of wellbeing. Social contacts in old age were mainly the result of investments earlier in life — a precautionary measure to avoid loneliness. Personality and a feeling of reciprocation were other important factors. Even for those whose level of social activity was low, the interactions they did engage in were of great importance for their feelings of self-worth.

One married couple reported being successful in old age, although they were anxious about how they would deal with the death of the other partner. They helped each other with mobility, because both had difficulty walking. He had a special relationship with a son and grandson who were struggling with mental illness. Both often visited him for moral support. His wife had good relations with another grandchild. Together, they had numerous social contacts, and were fully integrated into their local community.

Some of the over-85s were comfortably off, lived in nice houses surrounded by gardens, and had many social contacts, but still felt unsuccessful in old age. One woman spoke of how she no longer had contact with her daughter because of a conflict between them. That gave her a continuous feeling of loss, which tainted any experience of success in life.

A sense of wellbeing was measured by asking subjects to rate their quality of life as they experienced it, from 1 to 10. The average rating was 8. There were some participants who rated their quality of life as 1, but there were also a large number who responded with a score of 10. For most of the old people, wellbeing was more or less equivalent to feeling successful in life. It was not just important to adapt, but also to count your blessings, such as social contacts, and to focus on gains rather than losses.

It was noticeable that feelings of wellbeing were sometimes strongly associated with experiences earlier in life, or an anticipation of life after death. For example, an 85-year-old widower who had been married for 62 years said, 'Thank God I won't have to wait long until I see my wife again!' Despite his bereavement, having to move house and the turmoil associated with that, and increasing health problems, his sense of wellbeing remained great. He was intensely grateful for the years of happiness he had been able to share with his wife, and was looking forward to continuing them after his death.

One 85-year-old widow contrasted her positive present situation with a traumatic period in her youth. Her childhood had been dominated by feelings of fear and impotence, due to an incestuous relationship with her father. That dreadful time was followed by a happy marriage, which she looked back on with great satisfaction. Throughout her entire life, she had invested in social contacts with family and friends. Now, in old age, she enjoyed the affection of her children and grandchildren, and regarded herself as a happy person. Her present feeling of wellbeing was closely linked to the happy time of her marriage, and that enabled her to overcome the traumatic memories of her childhood. The physical limitations she experienced in old age were of secondary importance.

If 'successful ageing' is defined as a state of optimum physical, mental, and social functionality, then only around 10 per cent of the over-85s in Leiden would appear to fulfil those criteria. Gender, marital status, income, and education level appeared to have almost no influence on this result. According to the WHO definition, therefore, only a small fraction of our 85-year-olds were 'successful', and that percentage was even smaller among old people living in care facilities. That is really not an attractive outlook. However, this outcome was no great surprise to us researchers, since we knew that a large number of older people suffered from a range of chronic complaints and impairments. But we were pleasantly surprised by the fact that half the participants rated their own level of wellbeing as 'good to very good'. This is a case of the disability paradox, described earlier: when a person feels good, even though others tell them they are sick or impaired. The older people's own positive view of sickness and health was even more evident in the in-depth interviews: 22 out of the 27 interviewees described themselves, whether with a partner or alone, as successful and satisfied with their lives.

Further research was specifically aimed at finding out why older people who were not successful according to the WHO definition nevertheless regarded themselves as successful. First of all, it was important to identify the difference between independence and living with impairments. Several participants with a disability reported being able to manage well for themselves. The fact that this sometimes involved great difficulty, or was only possible with paid help from others, was irrelevant to them. Other participants had become dependent on others in their daily lives, because of ingrained gender roles rather than physical or mental impairment. For example, widowed men were often 'handicapped' in running a household, due to sheer inexperience. And widowed women were not uncommonly 'dependent' because their late husbands 'always sorted everything out for them'.

Finally, their immediate living situation was important in their ability to manage in day-to-day life: were there steps to climb, was a lift available, were there shops in the neighbourhood? Thus, some people were made dependent by hindrances in their environment.

Remarkably, some participants with dementia felt fine and, with support and supervision, were only slightly hampered in their daily lives by their memory complaints. Many older people distinguished sharply between physical and mental capacities or impairments. Body and mind were seen as important, but only insofar as they were necessary for the ability to function at the desired social level. Physical and mental impairments were seen as an inevitable loss due to old age. The opposite problem was more common: when older people were in good physical and mental shape, but did not feel happy, due to conflicts in their environment. Older people saw contacts with friends and family as a result of their own merits, as the payback for earlier social investments as parents, relatives, friends, neighbours, or co-workers. The quality of those contacts was seen as more important than the quantity, just as a large number of contacts in the present could not make up for one missing contact from the past.

The WHO definition reflects the way that researchers, doctors, and policymakers see health, and what they as professionals believe to be relevant when examining health issues. As we have seen, their way of seeing things is often at odds with the view that older people themselves take. Older people do not disagree with the various matters that outsiders deem important, but they attach a different level of importance to those factors.

In the WHO definition, for example, it is assumed that the different domains of day-to-day functioning and wellbeing are equally important; but, in the view of older people, there is a hierarchy. For most older people, it is social contacts and a feeling of wellbeing that determine their quality of life. The presence or absence of physical or mental impairments is less important. This is in accordance with what older people expect of the 'normal' ageing process — it is why they stress the importance of the ability to adapt as the key to achieving the higher goal of wellbeing. In order to feel successful, you have to be able to deal with loss and impairment, and integrate them into your life. Older people continuously refer to their own personal history and the immediate surroundings they live in. Those who have adapted better to their circumstances feel more successful.

An objective valuation by outsiders of wellbeing or the quality and quantity of social contacts does not form a basis for describing people as successful or unsuccessful. All that counts is how people assess these things themselves.

A RATING FOR LIFE

As a concept, 'quality of life' is newer than the WHO's 1948 definition of health. It was introduced by researchers as a way of contrasting people's feeling of wellbeing with the limited concept of illness as used by doctors and researchers.

There are many aspects that contribute to the quality of someone's life, and each of them should be included in any relevant questionnaire. However, this is where the inevitable problems begin. Which aspects need to be recorded, and how much weight should be attached to each of them? It is clear that old people put social relations first, but perhaps younger people attach more importance to other things, such as social standing, sex, or money. And it is different again for people living with illness. For example, questionnaires designed for patients with chronic lung disease must include a question about shortness of breath. Cancer patients must be asked about pain. These examples expose the real problem.

Doctors and researchers try as much as possible to put themselves in the place of the people they need to question, hoping that this will help them design good questionnaires to measure health, illness, and the quality of life. But the choices they make will always be the result of professional consideration, and not the choices of the interviewees themselves. Older people, younger people, and sick patients all find different things important in their lives. Some researchers therefore advocate letting people decide for themselves what aspects are important for their wellbeing. They differ not only between people, but also over the course of one person's life. What was important to someone when they were young, single, and at the start of their professional career may not necessarily be important to them in old age, when they already have a partner and perhaps children, too.

If the different aspects of wellbeing can vary so much from person to person, and can also change over the years, it would seem almost impossible to measure quality of life adequately using questionnaires. How can we afford the correct weighting to the different dimensions, to come to one result, one numerical value? The answer is that we can't, and so we should not even try. It makes more sense to take a different tack and ask the people in question directly about their quality of life, and request that they rate it numerically themselves. Each individual then determines what aspects are important for personal wellbeing, and delivers a well-considered assessment of them. If the rating is high, the researcher can then ask what aspects have contributed to this result, or what aspects are still deemed to be lacking. In fact, this is what we do automatically in our private or professional lives whenever we talk to our children, partner, parent, or patients. For example, a general practitioner or psychologist will start a conversation by saying, 'How do you feel?' If the answer is, 'Not good,' then the following question will be 'What is the matter?' or 'What can I do to help you?'

Asking someone directly how they feel about their own state of wellbeing is so obvious, it seems strange that we do not do it more often. Perhaps we don't want to hear the answer. Many of us are convinced when we are young or middle aged that the quality of life in old age is bad, without ever having actually asked any old people ourselves. Only too often do we say that all that matters is how you experience life yourself. How, then, can we claim to know how older people experience their existence?

Much research has been carried out in Europe into citizens' — not just old people's — sense of wellbeing. Usually, the question is something like: 'How satisfied are you, all things considered, with the life you lead?' The results are rather surprising. Quality of life receives the highest ratings in Denmark, Norway, and Switzerland, where people give it 8 points out of ten on average. It has long been known that the Danes are very happy people, although we do not know the exact reason for this. In the Netherlands, and in Australia, the average score is between 7 and 8; in the UK it is 7; and people in Italy give their lives a score of 6 on average. It is interesting to investigate the reasons for those national differences.

Ruut Veenhoven, the 'Professor of Happiness' from Rotterdam, has made a career of doing this. He has found that happiness bears only a limited relation to a country's gross national income, although large investments are necessary to maintain a high quality of life for citizens. However, it is not the case that more economic investment directly translates into happier people. There is, instead, a positive link between a well-functioning public administration and the quality of life. This suggests that a well-functioning state, based on the rule of law, and access to services and social provisions, are significant contributing factors to people's level of satisfaction with their lives.

In 2014, the UK's Legatum Institute established a Commission on Wellbeing and Policy to advance the policy debate on social wellbeing. Their aim is to give policymakers a greater understanding of how data on wellbeing can be used to improve public policy and to advance prosperity. The underlying principle is that prosperity is a more capacious idea than can be expressed by a purely material measure such as Gross Domestic Product.

Very importantly, perceptions of people in the same country but of different ages almost never differ. In Italy, citizens express the same level of satisfaction with their lives whether they are young or old; and in the Netherlands, as mentioned above, the average score is almost 8, irrespective of age category. Eight is also the rating given on average by the participants in the Leiden 85-Plus Study. The only difference between age categories is that people's sense of wellbeing is slightly lower when they are in their fifties. The fact that old people in the Netherlands rate their lives with a score of 8 means that most are able to cope with setbacks, illness, and impairment. The feeling of wellbeing increases slightly in extreme old age, perhaps because older people learn increasingly how to deal better with life.

The high rating does, however, go down in the last year before death, but, happily, that moment arrives ever later in our lives.

12

VITALITY

Modern care of old people takes too little account of the fact that everyone ages at different rates, and chronological age is not always the best indicator. A much more important guideline for care is provided by the four periods that occur in everybody's life — prevention, multimorbidity, frailty, and dependency — which together form the new 'Ages of Man and Woman'. If people are no longer allowed to be in control of their own lives, and individual responsibility is taken away from them, the consequences are disastrous for their vitality, wellbeing, and quality of life. This is why we are in dire need of a new approach to the care of old people.

On YouTube there is a video, entitled *100*, in which people from Amsterdam look into the camera and say their age out loud. It paints a wonderful portrait of their lives. In less than a minute, the viewer sees a parade of babies, children, teenagers, adults, and old people flash by. The youngsters are gloriously exuberant. They are followed by a multitude of people — some timid, some confident, some smiling radiantly, some with a pensive look on their face.

What this little film shows us is the ancient evolutionary programme that is coded in our genes. The programme is still appropriate for our development from birth to adulthood. After that, however, it is no longer adequate because there is simply no programme yet for the much longer lives we lead today. In the YouTube portraits, the people in their fifties have a slightly more sombre look in their eyes than the rest. That is perhaps because they are confronted with a faltering body, which is something nobody exactly looks forward to at that stage of their lives. It will take many more generations before our bodies are adapted by natural selection to the modern environment and our much-extended lives.

The people in the clip aged 75 and over all beam into the camera as they proudly announce their age. Perhaps people finally dare to show their true selves once they reach old age. Perhaps it is that they are finally able to do so, and that it becomes acceptable, since social constraints become more relaxed in old age. When people can be themselves, they will differ increasingly from one another, and that is certainly true of older people. Each person has his or her own trajectory of development; each person has accumulated damage; and, with a certain amount of adaptation, that makes each person an individual character in old age, some bearing their years with more dignity than others. When buildings and appliances show the marks of time, we find it attractive. 'You can see it's been well used.' Old people can impress us similarly with the way they cope with their lives. 'If only I could live to that age!' Then we don't mind the odd blemish here and there.

It is not only the speed with which the ageing process progresses that can vary between different people — some people age much earlier than others — but the signs by which ageing manifests itself can also differ widely. One person might lose cognitive function, but still look fantastic; another may be frail but still possess a razor-sharp mind. And then there is the question of how they lead their lives, of course.

Given that people age so individually, it is remarkable that we have so many regulations that are based on chronological age, as if we were all the same at the age of 55, 65, or 75. And we always talk about old people's care while tacitly assuming that we all mean the same thing by it. But the opposite is actually the case: there is no *one* way to care for old people. This is more than just quibbling about words. How are we to organise our society properly if we cannot even come up with a single, unambiguous definition of old people's care?

Although the onset of ageing and the course it takes can differ widely from person to person, some episodes in life that everyone goes through can be generally distinguished. The more sharply these can be defined, the better society will be able to accommodate our ever-longer lifespans. I divide the human ageing process into four periods: prevention, multi-morbidity, frailty, and dependency. Together, they make up the stages of a new 'Ages of Man and Woman' — a structured and chronological division of 'the rise and fall' of a person through life — but this time irrespective of chronological age, and highly adapted to the fortunate circumstances in which we now live. What has not changed is the underlying ageing mechanism that leads to illness, impairment, and disability, and finally to death.

In part, these new stages are the same as what the medical profession considers health, but the definition of them is determined to a large extent by the goal a person has in mind at a given moment, in keeping with the broad definition of health as used by old people themselves, and affords a prominent place to a sense of wellbeing. That feeling of wellbeing is, as described earlier, linked to the quality of a person's social relations and the environment he or she lives in. Medical problems do not play a large part in it.

THE NEW 'AGES OF MAN AND WOMAN'

Prevention is the first stage in the new 'Ages of Man and Woman'. It begins at, or even before, birth, and is by far the longest stage. Much is now known about the risk factors that accelerate the ageing process and about the factors that protect us and slow the process down. These are the ingredients for staying healthy, and, first and foremost, they are the responsibility of each individual: sitting a lot, eating a lot, and smoking, *versus* running, doing sport, and drinking in moderation. Despite all the information available, many people are still not sufficiently aware of the effect that good and bad habits have on them. And I am also referring here to the positive effects they can have on mood. Regular exercise creates a sense of wellbeing. Excessive alcohol consumption has the opposite effect.

In a world of plenty, many people are unable to keep themselves healthy into old age. They are targeted by people who want to influence their behaviour for their own profit. Young people are made dependent on alcohol, cigarettes, and fast food. Once they are hooked, such behaviour is difficult to reverse. This is precisely the aim of advertising that targets young people. It is an example of the way that the economic interests of a few take precedence over the many, and clash with a higher public interest: people staying healthy for longer. This is no different today than it was in the past. We now consider it a disgrace that children were forced to work for others' gain at the start of the Industrial Revolution. We must consider whether the situation today is any different. This conflict between citizens and business should give us sufficient reason to prioritise prevention in the public domain. To claim, as some do, that prevention is each individual's own responsibility is a blatant denial of the facts.

We are familiar with mother-and-child clinics from the start of our lives — from vaccination programmes, for example. At this stage in our development, everything is taken care of. This prevention in early childhood is a necessary start, but it does not go far enough. Staying alive in childhood is no longer the problem today. Looking after our bodies and minds as they develop to adulthood and beyond — that is the challenge of the modern age. Child obesity shows unequivocally that kids are eating the wrong things and taking too little exercise. This is an enormous public challenge. If the stairs in a building are difficult to find, of course you are going to take the lift.

If fresh vegetables are more expensive than fast food, of course that will attract those trying to feed a family on a tight budget.

The public agenda for our children's education is even more out of step. If we continue to insist on the rapid acquisition of specialist knowledge, we will be producing half-finished goods, from a social point of view. Nowadays, it is no longer sufficient to train children for a once-in-a-lifetime profession. We need to provide them with more skills than that. The most important thing is to make sure they realise that they are likely to live to be 100, and that they must take responsibility for organising their own life trajectory accordingly. We need a new attitude — 'Old is not lame, old is cool' — which they will also apply to their own later lives. We must prepare children for a lifetime of learning so that they can continue to participate in society on both a formal and informal level. This means that we must pay more attention to the acquisition of social skills. These are mainly learned in childhood, and must be sufficient to last a long lifetime.

There is no real agenda for preventive measures in old age, and so this needs to be developed from scratch. There are no, or almost no, official initiatives to promote good health among older people. Yes, people over 60 get a flu shot, but why are there no effective screening and intervention programmes for high blood pressure? Roughly half the cases of high blood pressure go undetected. And only half of those who are found to have high blood pressure receive treatment to normalise it. This result is nothing to be proud of. Doctors can now normalise anyone's blood pressure with medication. The fact that this does not happen is all the more distressing, since we now know that high blood pressure in middle age not only affects the blood vessels in the heart and kidneys, but also more than doubles the risk of developing dementia.

The obesity epidemic and sedentary lifestyles require serious action. Many older people think that, now they are of an advanced age, they have earned some rest. A sedentary lifestyle is tempting, but is very detrimental in the long run. Everyone loves food and comfort; that is an essential part of our evolutionary fitness programme. But in today's world of overabundance, we struggle with obesity, and it is clear that we must avoid such excesses.

It is not difficult to analyse the reasons for this undesirable situation. No one feels morally responsible for providing the necessary prevention agenda. This is compounded by the fact that many people earn a pretty penny from our unhealthy behaviour and the medical complications it leads to. A lot of health benefits are lost in this way, especially among the more economically disadvantaged sections of society. This lack of moral responsibility means we are still miles away from a situation where preventive measures help to promote health in a broad sense: not just the prevention of medical health problems, but also social problems.

One example can illustrate how important this is: older people with a small network of social contacts have a higher risk of mortality than smokers, even though smoking is seen as one of the greatest risk factors for illness and death. A healthy lifestyle includes investing in family, friends, and social contacts so as not to have to cope alone in old age. This is not only because not to do so jeopardises longevity, but also simply because it is not pleasant to grow old in isolation.

People will not be likely to take these preventive measures themselves, because there is no code for old-age care in our genes, and citizens are not sufficiently informed about such measures. Existing healthcare providers have no direct interest in introducing them, and simply ignore the looming challenge. Prevention must be shaped primarily by a sense of public responsibility. If we can translate our extra years of healthy life into paid or unpaid social participation, the putative problems of the ageing population will never arise.

The second stage of the new 'Ages of Man and Woman' is that of physical and mental complaints — due to tardy prevention. They eventually result in a visit to the doctor. He taps and listens, takes photos, and makes a diagnosis. On average, doctors are able to identify two or more chronic conditions in 65 per cent of 65-year-olds, and multi-morbidity — a fancy word for the co-occurrence of two or more chronic medical conditions in one person — is diagnosed in 85 per cent of 85-year-olds. These are almost exclusively complaints that develop as a result of the ageing process and are (as yet) irreversible. Unlike accidents, where the victim goes through a cycle of injury-surgery-rehabilitation-complete recovery, chronic conditions have no end. Lung function in people with emphysema deteriorates steadily; heart function diminishes gradually as a result of coronary atherosclerosis; and osteoporosis makes bones ever thinner. Once a chronic condition has set in, it is important to intervene quickly to keep any physical and cognitive complications at bay for as long as possible.

Our modern healthcare system is designed to deal with accidents and sudden illness in people who were more or less healthy before. It aims to address *one* problem as efficiently as possible, and then to discharge the patient from care. Specialist medicine is well able to treat single illnesses, greatly improving the prognosis for the patient. We have already seen the success of medical care in treating cardiovascular disease. Stomach ulcers are now almost a thing of the past, since the bacteria in the stomach wall that cause them can now be tackled with antibiotics. The prognosis for patients with Hodgkin's disease has improved drastically, thanks to a clever combination of radiation and chemotherapy. Modern inhalation therapy has turned asthma into a mild condition, in most cases. And the list of examples goes on. This is a real triumph of modern medical practice. But it is disenchanting to see that older patients with several concurrent conditions do not receive adequate treatment or support. Medicine and society respond to their needs inadequately, or not at all.

The problems are many and varied. Old people with a wide range of complaints are systematically ignored by scientific researchers. There is no good reason for this, other than the focus on young(er) people, which is taken for granted. This is a case of ageism — discrimination of people on the basis of age alone. The result is that we do not have sufficient scientific knowledge to provide the best care possible for most of the patients who present themselves to general practitioners and hospitals. Instead, specialist medicine leads to a jumble of different, uncoordinated actions. Most older people have to go through life with a list of appointments like a debutante's dance card, and a huge box of pills. Much more often than we would like, these treatment strategies and medications influence each other negatively. In such cases, less specialist medical care would lead to more health benefits. The need to coordinate and integrate different strands of patient care is increasing all the time, and this has led to a boom in the number of medical managers and care coordinators. However, since they are deployed separately for each specialist problem, some policymakers are now considering trying to turn the tide by coordinating the various coordinators and managers ...

More tools from the same box will not help with the complex problems of older people living with several (chronic) conditions. The crux of the matter is that the medical system of today is designed to cope with the health problems of the past. A new approach is necessary to enable us to treat today's complex patients properly. For the most part, the current divisions in medicine along the lines of individual organs and diseases must be broken down. General medical knowledge must be prioritised once more. The complex problems caused by different, concurrent conditions have become standard among old patients.

Who should feel called upon to tackle the problem of creating a new agenda for multi-morbidity? Everyone, in fact. Doctors, nurses, and paramedics will need to rethink the subject matter of their professions. Hospital managers will need to change the way departments and outpatient clinics are organised. Health insurers and policymakers must realise that a remuneration system based on specialist interventions is a hindrance to the development of a more generalist approach to the complex health problems of older people. The current system burdens patients with too many pre-treatment, treatment, and post-treatment trajectories, generating too many diagnostic assessments and extra treatments, as well as high costs to society. Governments must shift the current (financial) incentives towards helping to promote a change in care. Presently, the initiative lies far too much with the care providers, who still have too little interest in rethinking their offer. Older people must help force the necessary change, since the system will not change of its own accord. They feel sorely the lack of a general approach, they have moral rectitude on their side, and they will be the first to profit from such improvements.

The next stage in the new 'Ages of Man and Woman' occurs when several interacting disease processes render an old person 'frail'. This is the phase in life when minor setbacks can quickly snowball into a full-blown disaster. It is therefore important to identify frailty in older patients early, and then to implement medical and technical measures only with extreme caution. Prevention aims to stop disease developing, and, if it does occur, to use an appropriate method of treatment to avert any lasting complications, or to delay them for as long as possible. However, with frail old people, the primary aim is to preserve their ability to function in daily life.

A woman of 85 with an extensive medical history is diagnosed with cancer. The specialist proposes surgery and chemotherapy. A full recovery is no longer possible, but the treatment can slow the progress of the disease, so the patient agrees to it. Complications arise, and the woman spends the final months of her life in a hospital bed. The question is whether that is really what she wanted.

Doctors far too rarely offer the option of foregoing treatment. Far too often, they persist with a plan aimed at bringing about recovery, although they know that will never be the result. It is the responsibility of specialists to make it clear to patients when the expected effect of a medical intervention is minimal. Medics need to be far more aware that their primary role when treating frail old people is one of *support*. Many specialists believe they have no more part to play when a specific (medical) treatment is found to be impossible, or is rejected. But rehabilitation, adaptation, and support can enable a patient to live well with limitations and impairment, so that they can continue functioning well in their familiar environment for longer.

For frail old people in particular, medical action should be aimed at optimising their health in the broadest sense. This is easier when frail older people are allowed to set their own personal goals. This is where the general practitioner can play an important part, making explicit choices together with the patient. General practitioners can decide not to implement certain specialist interventions or treatments, and to sustain the wellbeing and functionality of the patient in other ways. General practitioners must become more like 'personal physicians' for their patients.

In Chapter 8, I illustrated how difficult it can be to identify frailty in patients. It will also therefore be difficult to prevent the unwanted side effects of medical interventions. But that does not relieve healthcare professionals of their obligation to pay this issue far more attention. Frail older people should not be treated as healthy adults who just happen to have grey hair, just as children should not be treated simply as miniature adults.

The fourth and final stage in the new 'Ages of Man and Woman' is dependency. This often concerns older people who have multiple debilitating conditions, or who have dementia. In this period of their lives, a wide network of informal carers and professional care workers becomes active. Older people can quickly lose control over their own lives in this stage. That is why it is so important that their family and friends are able to advocate for them. I call this a 'vicarious advocate'. It is the responsibility of older people to organise such an advocate themselves while they are still in a position to do so. This ensures that there will be no doubt about who will speak for them when they can no longer speak for themselves. Quality of life is the guiding principle; prolonging life retreats into the background. The goal now is to increase the old person's sense of wellbeing, rather than to postpone death.

Two-thirds of doctors in the Netherlands are of the opinion that people in the final stage of their life are often treated medically for longer than is desirable. Proposals to stop treatment go against everything doctors are taught during their training, when the entire focus is on saving lives. Doctors systematically overestimate a given treatment's chances of success. This gives patients the idea that modern medicine has a solution to everything. In this way, doctors and patients hold each other in the grip of an illusion that preserving life is always feasible. Medicine should provide care to people always with a view to the value it brings for them. A change of attitude is required when treating dependent older people.

Fortunately, it is possible in the Netherlands to speak openly about death. When there is agreement between doctor and patient, futile medical treatment can be discontinued promptly. Sometimes this brings the point of death closer, since prolongation of life is no longer being pursued at all costs. Because non-treatment means they suffer no side effects from medical interventions, patients experience a greater sense of wellbeing. In this final phase of life, the doctor must take a back seat, remaining available if symptoms such as pain, shortness of breath, or anxiety surface, to relieve them with consultation, support, or medication.

In a period of dependency, there is no longer a relation between wellbeing and specific illness. This stage requires primarily bespoke, humane care. The amount of care required by a dependent older person is naturally determined by the person's physical and mental capacities, by those closest to them, and by their immediate surroundings. Most people approaching the end of their life prefer to be at home, but admission to a care home is sometimes unavoidable. Such a move is often seen as a catastrophe by older people. The question is why. Obviously, they are afraid of the approaching end, but perhaps sometimes care homes fail to offer what older people would like. It sounds paradoxical, but in the phase of dependency, with the end of their lives in sight, people need more than ever to be in control of their own lives, or to know that those closest to them are in control.

Are the workers in the care home sufficiently able to listen to the residents and cater to their needs? I think professionals in long-term care think too often in terms of medical, technical, and legal issues, instead of making use of their own, hard-earned expertise. This leads them inadvertently to do things that are not to the benefit of (frail) older people.

OPTIMISM AND ZEST FOR LIFE

Within geriatric medicine, it is often said that we should not seek to add years onto life, but that we should add quality to years. That is a dubious statement. I believe everyone wants to live longer, provided that their quality of life remains good. However, people are primarily responsible for their own quality of life. Medicine can create the preconditions for this, by keeping people's bodies and brains in good form for longer, but, when it comes to it, people must find their own sources of satisfaction. They must ask themselves what they are living for, and what their ambitions are. If they have a clear view of those goals, they can strive to achieve them, even in old age. The manifestations of the ageing process then become less important.

Thus, life in old age does not differ substantially from life when we are young or middle aged. Everyone has personal ambitions that they would like to fulfil, and everyone encounters stumbling blocks, hurdles, and setbacks in pursuit of them. The skill is to be able to overcome or circumnavigate those hindrances. Luckily, that is precisely what we have evolved to do. If we gave up at the first sign of adversity, we would never have made it so far as a species. Only the nature of our ambitions, and the nature of the setbacks we face, change as we get older. While it is mainly our social position and material wellbeing that are important to us when we are young, we begin to cherish our social relations more when we are old, despite illness and impairment. Fortunately, by then we have learned a lot from life. This explains why most older people rate their quality of life as good or very good.

A follow-up study on morbidity was made of people

whose attitude to life was measured in middle age. When they developed an illness in old age, its progress was better among those who had a more optimistic outlook on life than did their peers in middle age. Following a heart attack, optimists were less likely to die, and more likely to recover more quickly and to achieve a better level of functioning after rehabilitation. It seems that having an optimistic attitude to life keeps you physically healthier. And the reverse also appears to be true — 'a healthy body houses an optimistic mind'. In the families that took part in the Leiden Longevity Study, the offspring of parents who reached a higher age than average had a more positive outlook on life than their partners.

Social scientists stress that vitality is an attribute that is important for achieving happiness in old age. Vitality is the ability to develop to full capacity, to get the most out of life. This will to make something of oneself depends on motivation and introspection: what do I want, and where am I heading? It requires a positive attitude, and the energy to invest in that outlook. However, people also need the resilience to cope with setbacks. A (slightly too) positive view of your own abilities — optimism — is necessary to achieve your ambitions successfully. For some people, this seems like an impossible task. Vitality — call it zest for life — is an inborn trait, and it has its basis in our evolutionary fitness programme. It matures during childhood and adolescence, and is honed by the experiences people gather through life. Just as with other biological phenomena, it is also a combined result of nature and nurture — the influence of your parents and your environment.

Vitality is also the ability to appreciate what is possible and what is not, a balancing act between your own abilities and the ability to accept help from others. There is nothing more frustrating than pursuing goals that are unachievable. You have to be able to adapt your ambitions to the prevailing circumstances, and then you can achieve the goals you have set yourself. Old people are better at doing this, because they have so much experience to draw on.

Apathy is the opposite of vitality, and is defined by psychiatrists as a lack of motivation caused by a disturbance in mental functions. It is similar to depression, but is not the same thing. Depression is characterised by the fact that it causes people to suffer from feelings of sadness. With apathy, it is not so much melancholy that plays havoc with the mind, as the example below will show. A married couple, well into their seventies, came to my surgery, and the wife was the first to speak. 'He used to have a busy job, but now he does nothing. He just sits on the couch, waiting. I'm worried there might be something seriously wrong with him.' I asked her husband what he thought about his wife's account. He looked at me, and answered hesitantly, 'I haven't got a problem. She thinks I've got a problem,' and he pointed to his wife. The man himself did not appear to be suffering in any way. But those around him were worried, since sometimes people's characters can change greatly.

Some teenagers are apathetic, and spend much of their time doing nothing at all. Others are so full of zest for life that it's tiring just watching them. This shows that vitality is not linked to age, and is also independent of gender, educational level, and socio-economic class. In any given group we find the extremes of vitality and apathy, and everything between. In the diagram on the opposite page, the biological process of ageing, progressing from left to right along the horizontal axis, is measured against a vital attitude to life, represented by the vertical axis. The biological process of ageing and vitality are largely independent of one another. This is not as obvious as it may seem. Often, people assume that the two axes are the same. People — doctors and researchers included — often mistakenly believe that slowing the ageing process and postponing the onset of illness and complications will directly benefit our sense of wellbeing. This is what we call medicalisation: viewing life exclusively from the standpoint of medicine and medical technology.

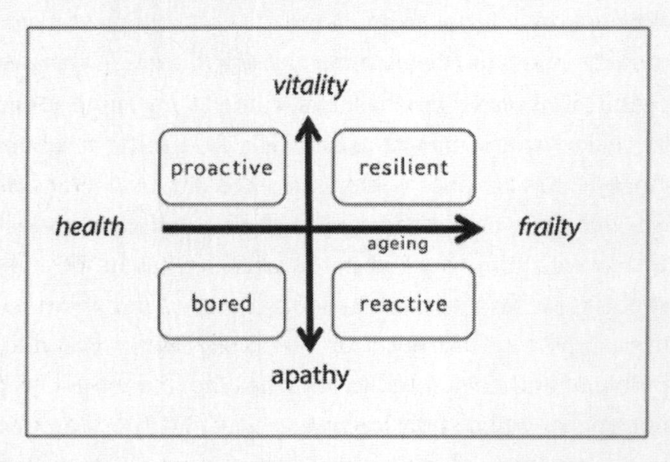

The horizontal axis representing the ageing process indicates the public responsibility of professionals to take the initiative for prevention and care. The vertical axis represents our own responsibility to strive for our goals in life. If older people consider social contacts to be of great importance, and want to exercise control over those contacts themselves, then we, as family and friends or as professionals, must play a supporting role. Quality of life, wellbeing, and happiness all result from vitality in life and an ability to take advantage of the possibilities available. The ageing process, health, illness, and impairment certainly have a part to play in this. They determine a person's level of physical and mental functionality, and an ability to manage day-to-day activities is contingent on them. But even if functionality is reduced, people are generally able to adapt — except in the final year of life, when, on average, satisfaction with life falls.

Our current thinking about increasing life expectancy is still too biased towards preventing and recovering from sickness and impairment, rather than concentrating on the day-to-day functioning and wellbeing of (frail) older people. This leads to a situation where the goals pursued by those involved in caring for old people do not always correspond to the ideas of those old people themselves. Professionals should not just concern themselves with the question of how we can stay healthy longer, but also with that of how they can stimulate older people to use their vitality in dealing with the problems they are saddled with due to the ageing process. Too little attention is paid to encouraging and supporting older people in remaining socially active, despite their sickness or impairment. This is what enables older people to maintain a high level of wellbeing. Friends, family, informal carers, and professionals need to stimulate people to keep on having dreams, nurturing their ambitions, and achieving realistic goals, despite their age and their reduced physical and mental functionality.

One example from my own neighbourhood is that of the old man who lives two doors down — the two of us first met when he was 85. He started working as a shipyard steelworker at an early age. He can talk animatedly about the time he worked for the union and took to the barricades to fight for a better deal for the workers. He was in his late fifties when he was sent into early retirement. He and his wife soon realised that they had to do something to give their life as a retired couple content and colour. He started taking lessons from a well-known local artist who was inspired by Zen Buddhism. My neighbour learned to paint; and when his hands became too unsteady for him to be able to continue painting, he took up an alternative art form. This led to the creation of a whole series of haiku poems. Some of his work was presented to the public in a small exhibition on his 89th birthday.

This new thinking about vitality and ageing requires a revolution in thought like that which took place in cultural anthropology. Until the fifties and sixties, anthropologists attempted to understand and explain people's behaviour from the point of view of their own norms and values. This is called the 'etic' approach. The view of the WHO on health is a good example of such an approach: a committee determines for other people what health is. This approach was abandoned by anthropologists, who began studying behaviour from the point of view of the people themselves. In this, the 'emic' approach, researchers throw off their own frame of reference, and adopt that of the people they are studying. This explanatory model, therefore, takes as its starting point the socio-cultural environment of the people who live in it, and the personal norms and values of the individual.

If we apply the emic approach to the care of older people, a meeting between a (professional) carer and the older person in question will begin with a dialogue between equals. The older person must be able to express his or her wishes and expectations. Professionals must be able to empathise with older people to understand their experience of the world and the problems they face. Only when that is clear can the caregiver be of service to the older person: 'What can I do to help you?' Naturally, this dialogue must be continued when medical treatment is under consideration. It is essential that such treatment is not only effective, but also feasible and acceptable to the patient. This entails the patient and doctor coming together to decide, which can lead to decisions being made that are in conflict with modern medical opinions, standards, or conventions. Research into this joint decision-making process shows that doctors and patients acting together often opt for a less-invasive treatment than that recommended in the guidelines. Not all older people's problems require medical treatment, but sometimes the doctor and patient may also decide to do more than usual.

Professional and informal caregivers need to take a new approach. Much more than now, older people's voices must be heard when it comes to the necessary restructuring of the modern healthcare system. The obvious starting point is the home, after which come family and friends, neighbours and, if necessary, the doctor and hospital. Currently, the process is completely the other way around. Enabling an older person to stay closer to home, in his or her immediate neighbourhood, we must do what we can to prevent a situation where all help for older people becomes medicalised. Many problems faced by older people require non-medical solutions, although illness and impairment will always be in the background. Those problems often have to do with feelings of wellbeing, and medicine cannot always provide an answer to such problems.

In public and professional debate, the question regularly arises of whether it is morally acceptable to demand of (frail) older people that they remain full of vitality; that they should set their own goals and achieve them themselves. If someone wants to sit apathetically on the couch all day, why shouldn't they be allowed to do that? On the other hand, we have to ask ourselves whether today's social order affords older people sufficient opportunity to maintain their vitality. Do older people have a worthy position in our society? Are they supported and valued enough? Guaranteeing mental and social vitality is a shared responsibility that requires good interaction. Everyone must come to realise how important it is to work on maintaining vitality in life. Older people should embrace the concept, because it is directly connected to their desire to take control of their own lives and to remain independent.

GREY IS NOT BLACK AND WHITE

The research report *Many Shades of Grey: ambitions of 55+* was published in 2013 to chart older people's own aspirations. The title is a reference to the fact that we see extremely positive or extremely negative images of ageing in the media, but never the 'grey' middle ground, which is where most older people find themselves. The research presents a cross-section of Dutch 55-to-85-year-olds. The report shows an optimistic future predicted for and by older people. They want to retain control over their lives, whether in the area of housing, working, care, or social contacts. Almost all want to be at the helm of their own lives, but when it comes down to it, they often feel powerless because existing public and private organisations will not allow it.

Increasing numbers of older people work actively and productively in the paid or voluntary sector. Social participation, even if it is unpaid, plays a large part in providing a sense of purpose to life, and in maintaining social contacts. But working in old age is not an obvious thing to do for many people. A large majority of over-55s who are still working take a predominantly positive view of the fact that they will soon be able to retire. However, there are also many who want to continue working, if it is possible under their own terms — for example, by going part-time or doing a different type of work. They are also willing to take a salary cut of up to 25 per cent. A quarter of those older people who no longer work say they would return if it were on their own terms. That proportion decreases, of course, with increasing age. Older people have a large sense of their own responsibility when it comes to their finances. Almost half those questioned felt in need of advice on how to deal with a possible drastic drop in their income, in order to pre-empt any problems that might entail.

As people get older, more of their relatives, friends, and acquaintances die. Despite this, the survey shows that older people are generally satisfied with their existing social contacts. The need for such contacts does not increase with age. Many older people argue that new social contacts must arise spontaneously and cannot be forced, especially in old age. The report shows the inherent strength of old people. The overwhelming majority have already thought about these issues, and say they feel able to cope alone with the loss of a partner or a loved one.

Almost all older people would like to take more responsibility for their own health. This is an argument for more appropriate care, and it would probably produce better results. But most older people do not really know how they can assume this responsibility for themselves. When the researchers made the idea more tangible by confronting the interviewees with a hypothetical health problem, the vast majority showed they were ready to take the initiative. They were keen to seek out information actively, and were prepared to adapt their lifestyle to the demands of the health problem. This also included them taking the necessary measurements — such as their blood pressure — themselves. Remarkably, they also displayed a desire to regulate their own use of medicines, as diabetic patients do with their insulin. The respondents' feeling of 'entitlement' to care increased with increasing age. Over the age of 75, the majority of Dutch people remain stuck in the old patterns they learned while growing up.

This survey shows that the majority of 55-to-85-year-olds are happy in their home, and would like to remain living there as long as possible. They have a wide variety of specific wishes for the future. In this regard, older people must take the initiative so that they are not forced to move at short notice as a result of increasing physical or financial problems. Only half of the older people questioned had thought about this.

Old people are caught in a dilemma, between 'Do I have to fend for myself?' and 'Am I able to fend for myself?' Older people need to start taking a different attitude to life, based on finding a balance between taking responsibility for themselves and making use of what public services can offer them.

13

THE NEW LIFE TRAJECTORY

Our average life expectancy has increased drastically, but society has not yet adapted to that development. We would do better to adjust our life trajectories to this increasing lifespan, which means working and remaining socially active for longer, and, by the same token, enjoying our retirement for longer than our parents and grandparents could. This will require a revolution in the way people, and society as a whole, think: more varied models of housing, more research into the types of disease that are characteristic of old age, more attention paid by carers to what people themselves want. Only then will older people find support in filling their longer lives with meaningful content.

At least once a year, Statistics Netherlands issues a press report on how life expectancy at birth has increased. The only thing that varies is their choice of words. Some years, it has 'increased slightly'; in other years, it's simply 'increased'; and in yet other years, it has 'risen considerably'. Not every year, it seems, is a record-breaker.

Almost every time such a press report comes out, I get a phone call from someone in the media. 'Mr Westendorp, what do you think about this?' And I answer, 'I don't think much about it, except that it's a good sign.' 'But ...' Usually a short pause follows. 'Surely this can't go on forever, can it?'

When we imagine ourselves getting old, we soon find our thoughts turning to our parents, grandparents, and other important people from the past. Using their stories and things we know about them, we try to visualise what it must be like to grow old, and to be old. But the past is the wrong direction to be looking in. We need to look to the future.

So have we no need for the stories from the past? Do we have no use for those who are now old? Of course we do. Old people and their life stories are inspiring, because they are not only directed towards the past, but look forward, too.

75 IS THE NEW 65

The lives of my grandparents, who were born around 1900 in Twente, followed a set pattern: primary school, then work, and then, for the lucky ones, a nice retirement. At the beginning of the Industrial Revolution, textile barons settled in that poverty-stricken area of the Netherlands (close to the German border), attracted by the abundance of cheap labour there. Cotton from America was woven into cloth in Twente and sold all over the world. Both my grandfathers were weavers — one working in Jordaan's mill, the other in the factory run by the Ten Hoopens. They were the main industrialist families in the region, and they provided people with work, pay, and shelter. The fat cats were regarded with both awe and contempt by the workers. It is noteworthy that both the bosses and the workers owed their rising life expectancy to the same industrial activities.

The families of both my sets of grandparents lived on the same dirt road, from where they watched the twentieth century develop. They saw the first cars, the first war, the second war, and the increasing prosperity from the nineteen-fifties onwards. First they got streetlights, then the telephone, and then television. They worked hard for the boss. If things went well, that meant 25, 40, or 50 years of 'faithful' service. At least that's what was on the certificates they were awarded to commemorate those anniversary years.

One of my grandfathers stopped working in 1965. He was 65, and had a good ten years of retirement ahead of him as a reward. On his last day, he was picked up at home by the company car — the greatest possible honour for an ordinary worker — and in Mr Jordaan's office he was awarded a commemorative ribbon. Sixty-five years old, retirement, a milestone: he had made it. In the evening, there was a party. I can remember the terrace full of people and hustle-and-bustle. The village band came marching down the dirt road in full regalia. The start of my grandfather's retirement was accompanied by marching music. My grandparents had led a frugal life. Like all weavers, they owned a small parcel of land and some small livestock to cover their basic needs. Saving money for their retirement was not an option. But, unlike *their* fathers and mothers, my grandparents did not lapse into poverty and financial dependency. They, after all, had their state pensions.

The Netherlands was already rich in 1959, but when the Groningen gas field was discovered, there was a new source of cash to fill the nation's coffers. This money was used to set up the welfare state. In the nineteen-sixties and nineteen-seventies, counterculture activists and feminists, spurred on by the Paris student revolts, forced Dutch society to change its perspective on life. The term 'senior' was introduced as a neutral, non-discriminatory designation for people over 55. The principle of non-discrimination was to prove a vain hope, though, as 55 soon became the age at which seniors were force to retreat from the world of work, or they were lured away by early-retirement schemes. The introduction of state pensions meant that poverty in old age had become a thing of the past, and far greater possibilities were opened up by the economic prosperity of the post-war Netherlands and its internationally unprecedented level of investment in the pension system. You could now retire early on full pay! Social pressure from the 'baby boomers' meant that older people were expected to leave the socio-economic stage at an ever-younger age to make room for that generation. Around the turn of the millennium, the de facto age of retirement in the Netherlands had dropped to well below 60 years.

Thus, the increased birth rate and the country's economic prosperity after World War Two led to a change in life trajectories that, it turns out, is not desirable. Early retirement has led to a drastic reduction in the length of time that men and women are seen as fully fledged members of society. Our 'social life expectancy' has fallen radically, and people are declared old at an ever-younger age. Fifty-five has become the new 65, while our biological life expectancy keeps on increasing. Dutch people are also staying fit for longer and longer; life expectancy without impairment has never been so high.

It is patently obvious that the period in which we are active contributors to the social good, whether working or not, whether in a paid or unpaid capacity, will have to increase rapidly. In short, we must start working longer again, and remain active participants in society, while still looking forward to a 'retirement' that is longer than that of our parents and grandparents.

This extended life trajectory requires us to reset a number of age-related social milestones. First of all, we must banish the term 'senior' as a gateway to retirement. That label can easily lead to a retraction of full membership of society for older people, and has recently led some individuals to demand unrealistic rights. We are now living longer than the number of years we pay into the pension system. Our problems funding the pension system can only be solved by rapidly adapting the pensionable age to our remaining life expectancy. This means something other than the gradual shift towards 67 that we are seeing at the moment. In view of the continuous increase in life expectancy in old age, it is realistic to move towards a pensionable age of 75. Seventy-five is the new 65!

WHO IS RESPONSIBLE FOR WHAT?

For the baby boomers, there is something bitter about the redesign of our lives as described in the previous section. After all, in the wave of economic prosperity, it was the post-war generation that contributed considerably to the shortening of working lives. Older people had to make way for the younger generation, and many of them went into early retirement. Now the baby boomers are getting old themselves and are being held to account. Not that they are short of money themselves — older people have never had as much income and so many assets as they have now — but they are being held responsible for the great financial burden that their retirement imposes on society. Many younger working people are no longer willing to bear that burden. The roles have shifted: just as the baby boomers in their time fought against older people in established positions, they now find themselves besieged by the younger generation.

In 2012, a book entitled *Mother, When Will You Finally Die?* caused a furore in the German media. The title was a deliberate provocation aimed at stirring public debate. Germany has relatively few young people and many older people, and this puts pressure on the relationship between the generations. In addition, in Germany it is more common than in the Netherlands for children to care for their elderly parents. When parents become dependent on long-term care in old age, and are not in a position to pay for it themselves, their children are expected to provide that care in the first instance. This responsibility is even regulated by German law. Martina Rosenberg, the author of the book in question, and a daughter of parents with dementia, describes her experiences of caring for them, and the devastating effect it had on her career and her own health.

The commotion caused by the book crossed the border into the Netherlands. The Dutch population is not as aged as Germany's, and the country appears to have escaped population shrinkage — for now, at least. But the rising cost of medical and long-term care means that the problem described in that book is relevant to the Netherlands, too. Already everyone in the country invests an average of 25 per cent of their wages in future care; that is the sum of all the taxes, premiums, and direct payments. If policies do not change, that proportion is set to rise to 50 per cent of income in the coming decades. Roughly a quarter of those costs relate to long-term care. The race is on to find alternative ways of organising long-term care, in order to keep the state budget balanced, to stop public spending from spiralling out of control, and — most importantly — to guarantee that money is available for other things such as infrastructure projects, education, and culture.

To rein in care costs, the conditions for claiming for long-term professional care were tightened in the Netherlands. Among other things, the number of places in residential institutions was reduced. However, a shift is also taking place from formal to informal care. When the government steps back, the question is who is to assume responsibility for the care of dependent older people. A glance at Dutch social history shows us that the three-generational family, where grandparents live with their children and grandchildren, was not very common in the country. In the Poor Law of 1854 it was established that relief for the poor, which included care of older people, was primarily the responsibility of the Church, as well as some private bodies. It was not until the Poor Law of 1912 that families had to take responsibility for their poor members first, before they were able to apply to the Church and other institutions for help. Although this was referred to as a 'moral obligation', this was clearly a cost-cutting measure, and enabled the government partly to withdraw from this responsibility. Caring for their older members was a challenge for many young families, which often had problems making ends meet. There was a lot of resistance to the law at the time, as evidenced by the many applications for mediation made to the 'Committee on Maintenance Obligations'. Relief for older people came with the introduction of state pensions in 1957. Unlike in Germany, families in the Netherlands were finally absolved of the financial responsibility of caring for older people with the advent of the Social Assistance Law.

In the post-war period of reconstruction, there was a rush to set up 'old people's homes'. There was a dire shortage of housing, and older people had to make room for the many large young families. In addition, the construction of separate old people's homes was a reaction to the lack of public provisions for dependent old people, and to the badly run private facilities. Thanks in part to the 'inexhaustible' income from natural gas, the government was able to assume a number of responsibilities that had formerly been left to private initiatives. Thus, in the nineteen-sixties and nineteen-seventies, complexes were built all over the Netherlands to offer institutional care and nursing for older people on a massive scale. The huge growth in the number of facilities was in part a result of competition between various distinct religious organisations. They made it a point of honour to establish residential-care facilities for their 'own population'. This is partly the reason why institutional care is much more prevalent in the Netherlands than in other developed countries.

Long-term care has now become extremely medicalised, not only in the Netherlands. Old people's homes have become care homes, and care homes have become nursing homes. That is fitting and natural for certain groups of (older) people with specific, unpredictable care needs, such as those with dementia. But it is less reasonable for old people with normal, foreseeable care needs. All those involved are in the thrall of the three-headed spectre of segregation, institutionalisation, and medicalisation. This must change, and that change is possible, because older people want independence and control, and policymakers want to keep rapidly burgeoning costs in check.

A major intervention in the current care system is needed, because a reallocation of costs, obligations, and responsibilities cannot be put off any longer. The state must introduce stricter criteria for determining care needs, to scale back public responsibility. Overall costs will not fall in the near future, as an increasing number of older people will become dependent on publicly financed care. That is not because we are living longer, but because the baby-boom generation will get old over the next twenty-five years. Stricter criteria for determining care needs will not, of course, reduce the needs of individual old people themselves, or their families and informal caregivers. Negative reactions to this redistribution of responsibilities are to be expected from individual citizens: 'And am I going to have to pay for all this?' is a predictable reaction. Just like the process of dealing with any other major loss, feelings of denial, anger, and dejection will alternate with one another. However, there is as yet no sign of this sorely needed new balance between public and private responsibility.

The younger generations are simply not dealing with the facts of the new life trajectory. They still hope to grow old in good health and then die suddenly. That is why they cannot conceive of the problems we are facing today. Children who care for their old parents often find it too much to cope with. They already have enough on their plates dealing with their busy modern lives, and feel unable to take on any more tasks. There is a constant process of negotiation between social partners over contribution levels, pension rights, and the extent of public provisions. But there has been no success as yet in regulating the financial responsibilities between the generations. Most working people have unrealistic expectations about the time when they will be able to go into retirement, and believe they will be able to remain independent over the long term. Many older and younger people feel despondent and insecure because they do not see the facts clearly. More education, information, and public discussion are needed to change this state of affairs.

I sometimes wonder whether people in the West are fully aware of how good their situation is. The welfare state is too often taken for granted. Being so used to the situation, older people can end up placing too much faith in institutional solutions, and not using their own powers of organisation. This is all the more remarkable in view of the fact that old people and the associations that represent them have made issues like 'control over one's own life' and 'remaining independent' their core values. If people have been responsible for organising their own lives throughout their adulthood, it stands to reason that they should also be the first to take responsibility for the content of the latter part of their lives. This must be possible without it being a threat to older people's right to exist. However, social responsibility, and in particular that of younger generations, is also inherent in this call to older people. Should we follow the German lead, and transfer the responsibility of care for older people onto their children? The initial and widespread reaction to this is one of personal repugnance, especially among young people themselves. 'Who wants to give their parents a bath?' Never! The bond between children and their parents is a deep well of interpersonal responsibility, but there are limits even to that. Wiping bottoms is probably beyond that limit.

It is extremely important to keep the discussion about intergenerational responsibility going. The debate is currently hindered by older people with too many demands, and younger people who feel overwhelmed by their obligations. The discussion is too often poisoned by stereotyped images of pensioners living a life of luxury, or of frail old people who are only to be pitied. Sometimes older people bristling with vitality and in full control of their lives are looked on with pity, as people ponder how much longer it will last. But there is nothing which says that old people do not have the right to make their own plans, especially when a 65-year-old can expect to live another twenty years, fourteen of them without impairment. It is gratifying to make plans and then realise them, to be the architect of your own future.

The worlds of science, government, and business must join hands to bring about (social) innovation. There is a need, for example, for a wide range of diverse housing options for older people. This means more than just luxury housing with top-level services, but also housing solutions where people on low incomes are also ensured sufficient care. Until now, that need has been met with a more or less standard range of care and nursing homes. Cooperation will be necessary to take advantage of opportunities that have so far been missed.

The labour market is also in need of a major overhaul so that older people can demand the right to work, if that is what they want, or if they are forced to do so by lack of personal finances. The system of full-time work until the age of 65, and then nothing, offers too few prospects for the future. Additional training and retraining measures, part-time work, and career downgrading can provide a way for older people to remain in paid work.

In medicine, research and innovation must be focussed on prevention and preserving functionality in old age. Professional caregivers must support older people in their efforts to preserve health and day-to-day functioning, rather than dictating to older people and taking away control over their lives. But the most important task for the future is to support older people in utilising their organisational capacities themselves. How can society give them the power to make their own plans and realise them themselves? It is our moral duty to guarantee frail and dependent old people the right to a proper place in society, just as we have done for young adults and people with disabilities.

Time is running out. In Japan, finance minister Aso

called on terminally ill patients to 'Hurry up and die', so as not to be a burden on state finances. In Germany, there are repeated reports of children who buy their care-dependent parents a 'one-way ticket' to Ukraine, Slovakia, or Thailand to be looked after there. Luckily, there has been no talk of such things in the Netherlands. But these incidents from Japan and Germany make it painfully clear that older people who are heavily reliant on the state or their family are in a vulnerable position. We should be able to choose for ourselves who wipes our bottom later in life.

A BIRTHDAY RHYME

'What do you say to your father on his 70th birthday?' a close colleague asked me recently. 'You make up a rhyme for him, as usual,' I said, referring to the Dutch birthday tradition. 'It doesn't have to be pompous, like in the old days, when you would have to bow deeply as you proclaimed the first lines. A little teasing can't hurt. It's more appropriate for the new relationship between the generations. But you'd better be a bit careful. At their age, our mothers and fathers are easily offended.'

Lesson number one for the birthday ditty: older people feel they are special, but they are not. After all, the chances of reaching the age of 70 have now risen to 80 per cent.

I remember a phone conversation I had with my own mother a few years ago. I had argued her into a corner, and she snapped, 'I'm 74 years old!' to let me know she was superior and I should just shut up. It only got worse when I replied that she was only 'born yesterday', as far as her age was concerned, and added that reaching her age was nothing special. That didn't go down well. End of conversation ... The danger is that older people only ever speak to other older people; that they will simply end up parroting each other's opinions, and grow blind to the fact that they really aren't anything out of the ordinary. So I advised my colleague to tell his father that he was nothing special.

That is not an easy message to deliver, so I suggested he take the role of court fool. The fool was the embodiment of truth and reason at the royal court. He was permitted to say what was forbidden to others. He held up a mirror to the king, reflecting his mistakes and his weaknesses, and served his master in this independent way. In short, the aim was to dethrone my colleague's father in his anniversary year; to take the 'look-at-me-I'm-70!' wind out of his sails in one fell swoop.

'Lesson two,' I continued, 'is to try and draw him out a bit. Lines like "Father, what are you going to do next? What are your plans?" are a good way of doing that. A lot of older people are shocked when you ask them this. They look at you in confusion and say, "How do you mean?" But you have to say it to him. Now he's turned 70, his remaining life expectancy has risen from around ten years in 1943 — when he was born — to almost fifteen years now. Especially if he didn't smoke, or gave up in time.' But, I impressed on my colleague, you can't blame him for being shocked at the question. He is but a child of his time, after all.

And that led me to lesson three: when you've got your father slightly on the back foot, finish off your ditty with something along the lines of, 'Dad, think like a Maoist. They would always make at least three five-year plans ahead of time. Come up with a really cool plan. Surprise us. Surprise yourself. Then you'll have something to tell us about when you're 85.'

A PRESCRIPTION FOR THE FUTURE

'Is there a prescription for growing old without being old?' I am often asked this question, and it always makes me feel a little uncomfortable. I do not have a simple prescription that would prevent, solve, or alleviate the impairments associated with old age. There is no miracle cure, and it is not likely that there ever will be. The causes of the damage that arises due to the ageing process are simply too complex and multifaceted.

How different that is from the message we get from the anti-ageing industry. It makes convenient use of what we all want to hear: that the ageing process can be prevented, or at least slowed, by using the product being touted.

There are countless special diets that 'gurus' claim can keep you young. There is, however, no evidence for their claims. Fortunately for us, the quality of our food is better than ever before, in the sense that it is not contaminated, toxic, or rotten. Many harmful ingredients have been banned by law. The problem we face today is that we consume too much of some substances and too little of others. Our food often contains too much salt, fat, or sugar, and many of us do not eat enough fruit and vegetables.

Nutrition is a complex matter, and experts have great difficulty in formulating precise guidelines. Their advice is highly subject to the vagaries of fashion. Remember the nineteen-eighties and nineteen-nineties, when we were confidently informed that every egg we ate took us one step closer to a heart attack. Now we are allowed to enjoy an egg for breakfast every day. Twenty years ago it was impossible to get hold of butter or bacon in an American hotel; whole grains in all shapes and sizes were the standard. Now, you see Americans tucking into their bacon and eggs every morning, and carbohydrates are the new bad guys. For now, the best advice seems to be to eat a varied diet that includes a little bit of everything. 'Colourful eating' is the new buzzword.

Should we then just sit round idly, waiting until it's time for our final curtain call? Of course not. We can change our lifestyle at any time if it is negatively affecting our health. It is never too early and never too late to start. So why does it go wrong so often? Why are we too fat, why do we drink too much, cycle too little, and stubbornly refuse to give up smoking? Presumably, people underestimate the extent to which the environment we live and work in impacts on us. Advertisers constantly send out massive stimuli to influence our behaviour in a particular direction — often the wrong one. Take electric bikes, for example, the new must-have gadget among older people, which have been promoted massively in many developed countries. But why not just keep cycling the normal way, pedalling under your own power until you get tired? The adverts have the answer: 'Why tire yourself out pedalling when you can buy an e-bike?' Unless you use an e-bike instead of taking the car, it stops you from doing what you should: taking active exercise. Riding a real bike requires exertion, but you get a lot in return: health benefits from the positive stress on your heart, lungs, muscles, and joints, as well as an adrenaline rush. If there is one thing that researchers agree on, it is that a sedentary, inactive lifestyle should be avoided at all costs.

It is becoming increasingly clear to researchers that we can influence our behaviour positively by making small changes to our environment. Positive stimuli play a significant role in this. The success of Weight Watchers, for example, is due to the fact that it is not based on punishing its members for being too fat, but on rewarding them for reaching their target weight. Such positive stimuli can easily be integrated into our daily lives, making good habits easier to keep up, and increasing the overall beneficial effects. Research has shown, for example, that employees are willing to take the stairs when the lift is blocked. Scandinavia is the frontrunner when it comes to making such changes in the workplace. High desks and conference tables are installed in offices, forcing workers to stand, and so break the cycle of a sedentary existence. This also results in shorter, more efficient meetings, and lowers employees' risk of developing cardiovascular disease.

Another example is what happens in a railway station. When the escalator is broken, everyone walks up the steps. If the escalator is working, nobody bothers to climb the stairs themselves. But when some station steps in Stockholm were transformed into giant piano keys that played a note when stepped on, everyone suddenly used the stairs rather than the escalator. This is what researchers call 'serious gaming': actively enticing people to engage in healthy behaviour by challenging and rewarding them.

The extraordinary growth of the Internet and computer technology means there is still a world of benefits to be gained using such creative ideas to promote healthy behaviour in both young and old.

The most important key to health gains lies in our everyday surroundings. That is why doctors and researchers need to develop new public-health policies. They must use the same weapons as advertisers and promoters do to influence our behaviour. If you are confronted with vegetables at eye level when you open the fridge door, you are more likely to eat them than if they are hidden in the lower drawer. If you see cans, processed meat, and cheese, you will grab them first. These are precisely the tricks that supermarkets use to entice us to buy certain products.

A similar effect applies to the utensils we eat with. Almost without our noticing, the plates we eat from have increased in size over the past few decades. If you happen to come across your grandmother's dinner set, it is astonishing to see the 'saucers' they used as plates. Our wine glasses have become downright enormous. Experiments have been carried out in which large plates were replaced by smaller crockery, while nothing else was changed. The result was the people ate less and lost weight. People also eat less when they use smaller cutlery. If you have to do more spooning, you eat less soup. Not out of laziness, but because you feel full more quickly.

We have only a very patchy idea of the stimuli in our surroundings that prompt us to engage in healthy behaviour, and those that lead us down the wrong path. Much research still needs to be done into this area. In many households, for example, the dining table has 'disappeared', and the kitchen is left unused. Fast food, the couch, and the TV have taken their place, especially in families with an obesity problem. How would eating patterns change if the dining table were returned to use?

When it comes down to it, there is no way to prevent impairment to our bodies and brains. In the preceding chapters, I have shown the possibilities currently offered by medical technology and biomedicine for prolonging our functional life. Whatever impairments we develop after the age of 50, they are increasingly repairable. Medical technicians are chomping at the bit to come up with new hips, valves, and lenses. They are busy developing implantable devices to support a faltering heart, and perhaps one day even replace it. Fatal bleeding can be prevented by putting ingenious plastic pipes into dilated blood vessels. It will not be long before we all have a new little microphone implanted in our ear as soon as the built-in one we were born with fails due ageing.

Medical biologists are also busy. We are likely one day to solve the riddle of the hydra — endless regeneration from stem cells. The first step has already been taken. Patients whose bone marrow has been depleted by disease or chemotherapy can already be given new bone marrow made from embryonic stem cells taken from umbilical-cord blood. Scientists are also working on reconstructing the retinas of those who have lost theirs due to diabetes, for example. The possibility of rebuilding damaged gut tissue using deep-layer stem cells shines on the distant horizon.

This may all sound like science fiction, but all such solutions sound unlikely until a breakthrough comes. Before penicillin was available, half of those who contracted pneumonia died. No one thought a solution to the problem was in sight. This is unimaginable to us today. Medical breakthroughs will continue to happen. But no one knows exactly when.

There comes a moment for everyone when life is almost over and there is no point in tinkering with it any longer. Some people say they want to avoid that decline. They want to take action to allow them to end their lives at a moment of their own choosing. It is noticeable that more and more people are expressing such thoughts at an ever-younger age. This is because they prize an 'unrestricted' existence, and do not want their lives to be slowly 'skimmed off'; and it is also because old people are considered undesirable in our current culture.

But are the physical and mental limitations of blind or deaf people, for example, really a disaster always to be prevented? Experience teaches us that most older people do not (want to) take that decision when the time comes. Unfortunately, loved ones and professionals do not always react correctly to increasing impairment. All too often, physical dependency is seen as sufficient reason to write someone off as non-active and to take over control of their lives. This robs a person of their independence and dignity.

In the final, vulnerable stage of their lives, in particular, vitality is necessary if older people are to make their day-to-day life enjoyable. This is all the more true of people who are less able to look after themselves, for whatever reason. In that case, they need someone — a loved one or a professional — to help them fulfil their wishes and meet their needs.

Fortunately, most older people are able to run their own lives. Their failing bodies are then 'merely' flaws they can easily live with. Since many older people no longer see their physical, and often also their mental, limitations as a burden, they feel 'unbound'.

The real answer to the question of how to grow old without being old lies in our own social and psychological flexibility. Old people show us that. Again and again, I am struck by older people who have managed to retain their vitality and sense of wellbeing, despite impairment and limitations.

Aafje is just such a person. At 96, she is a phenomenon. She can often be seen riding around the neighbourhood where I live on her conspicuous scooter, which she handles with skill. You can see her outside the nursing home, or sipping an espresso on the café terrace. At the baker's she chats with the other customers in the queue. On many occasions I have seen her whizz by in a taxi, a woman on a mission: to get her hair done.

I have also met Aafje a couple of times in my professional capacity. Not that she came to my surgery with a medical complaint; she came to tell people how best to get old. At the Leyden Academy, we organise meetings with older people as part of our course for care professionals. These carers often focus on organisation rather than on the old people they care for. Aafje was one of the 'experience experts' invited to come and tell the course participants how old people can deal with loss, illness, and impairment, and a stiff and leaky body, and still run their own lives and retain their dignity. She had lost her husband and had had to move out of her 'gorgeous' house. She was no longer able to live alone at home, despite the army of helpers she had drummed up. Dressing and undressing — let alone getting in and out of the shower — were no longer possible without help. However, with her mischievous smile and her freshly coiffed hair, Aafje made a fragile but unforgettable impression. 'Let it go,' she told us. 'You have to let it all go.'

Acknowledgements

This book is a personal account of my journey through the field of gerontology and geriatric medicine. I have worked in the speciality since 1997, as an internist, scientific researcher and lecturer at the Leiden University Medical Centre (LUMC), Leiden University (LU), and, from 2008 to 2014, as the director of the Leyden Academy on Vitality and Ageing (LAVA). During that time, I have grown older myself. The voice of the person in their fifties heard so often in this book is representative, probably far more than I originally had in mind, of my own quest to find a way to cope with the ageing process. That means that my views do not necessarily always reflect those of the institutions with which I am professionally connected.

When I first entered the field of gerontology and geriatrics, I was largely unfamiliar with the subject matter. Although I had seen a great many old people in my capacity as a general internist, I had little inkling of why or how we age. The insights presented in this book are the fruits of seventeen years of working with many other doctors' and scientists' ideas and thoughts, and it is with great appreciation that I have depended on their work. Below are some references to sources from which I have so gratefully drawn.

Both scientific endeavour and medical innovation typically

work by interpreting and building on the discoveries and ideas of others. This method is described in many wonderful books. One such work is:

Friedman, M. & G.W. Friedland (2000). *Medicine's 10 Greatest Discoveries*. Yale: Yale University Press.

Knowledge and insight is one thing. Putting them down on paper is another altogether. In this respect I am much indebted to Silvia Zwaaneveldt, who taught me how to spell, to Jan Vandenbroucke, who taught me how to write, and, finally, to Ine Soepnel, who taught me how to put a book together.

Further Reading

AN EXPLOSION OF LIFE

Dr Robert Butler (1927–2010) was the first director of the influential National Institute on Aging (NIA, USA). He is best known for his 1975 book *Why Survive? Being Old in America*, in which he exposed the marginalisation of older people. For a biography of this champion of older people's rights, see:

Achenbaum, W.A. (2013). *Robert N. Butler, MD: visionary of healthy aging*. New York: Columbia University Press.

Chapter One
THE RHYTHM OF LIFE

An accumulation of damage

For a further introduction to the ageing process, I recommend the website of the biologist Dr João Pedro de Magalhães (Heswall, Wirral, UK), a source of inspiration with references to further information for beginners and advanced students:

Introduction to the ageing process: http//www.senescence.info/.

In his now-seminal lecture, given in London in 1951, the British zoologist and later Nobel Prize-winner (Medicine, 1960) Peter Medawar (1915–1987) describes the principle of the accumulation of damage as the basis of the ageing process:

Medawar, P.B. (1952). *An Unsolved Problem of Biology*. London: H.K. Lewis & Co.

All for the next generation

The conclusion of this section is that ageing 'can' only occur because natural selection is the dominant force at the beginning of the human life cycle. The evolutionary biologist Professor Stephen Stearns (Yale, USA) is one of the world's greatest thinkers on the different phases of the (human) life cycle. His book is considered a standard reference work:

Stearns, S. (1992). *The Evolution of Life Histories*. New York: Oxford University Press.

A more recent book on the topic is:

Thomas Flatt and Andreas Heyland (Eds.) (2011). *Mechanisms of Life History Evolution: the genetics and physiology of life history traits and trade-offs*. Oxford: Oxford University Press.

The biologist Professor Steve Austad (San Antonio, USA) is one of the great names in the field of ageing. He stresses that (human) ageing is not just a phenomenon that (today) occurs under favourable conditions, but which occurred (in the past) under unfavourable conditions. He substantiates this with an overview of recent data gathered by biologists on many different animal species:

Nussey, D.H., H. Froy, et al. (2013). 'Senescence in natural populations of animals: widespread evidence and its implications for bio-gerontology.' *Ageing Research Reviews*, 12: pp. 214–25.

Rites of passage

The best description of the course of our life trajectories in different periods of human history can be found in a richly illustrated work put together by the historian Professor Pat Thane (London, UK). One of the most important points she makes is that the social status of old people differs from one historical period to another, and from culture to culture, and that it declines rapidly in times of economic depression. She postulates that it is a misconception that older people always enjoyed a high social status in the past:

Thane, P. (Ed.) (2005). *The Long History of Old Age*. London: Thames & Hudson.

The second issue dealt with in this section is the genetic basis for our life history. In it, I describe the different forms of the nematode (roundworm) *C. elegans*. Those forms cover a range — selective constraints — that can include environmentally triggered plasticity. The evolutionary biologist Paul Brakefield (Cambridge, UK), is the master of so-called 'evo-devo' (*evolution of development*) theory:

Brakefield, P.M. (2006). 'Evo-devo and constraints on selection.' *Trends in Ecology & Evolution*, 2: pp. 362–68.

Chapter Two
ETERNAL LIFE

Damage and repair

The rate of ageing is the result of an accumulation of damage on the one hand, and maintenance and repair on the other. These are two sides of the same coin. Below are references to the work in which it was observed that hydras do not age thanks to the presence of 'totipotent' stem cells:

Martínez, D.E. (1998). 'Mortality patterns suggest lack

of senescence in hydra.' *Experimental Gerontology*, 33: pp. 217–25

Galliot, B. (2012). 'Hydra, a fruitful model system for 270 years.' *International Journal of Developmental Biology*, 56: pp. 411–23.

To be impressed by the enormous variation in lifespan — including the non-senescent hydra — see:

Jones, O.R., A. Scheuerlein, et al. (2014). Diversity of ageing across the tree of life. *Nature* 505:169–74.

Longevity in families

In the past ten to twenty years, researchers have developed various methods to help them discover the genetic basis for longevity. The first method is to look at people who are over a hundred years old — centenarians. The gerontologist Professor Claudio Franceschi (Bologna, Italy) has the most experience in this area of research. The second method is pursued by the endocrinologist Professor Nir Barzilai (New York, USA), who included *only* long-lived Ashkenazy Jews and their offspring in his study. The third method involves gathering information on families in which *several* nonagenarian siblings are still alive, and collecting data concerning their children and their partners. This last method increases the chance that the longevity of the participants is due to hereditory factors, and is followed by the Leiden Longevity Study, initiated by the biologist Professor Eline Slagboom (Leiden) and myself:

Cevenini, E., L. Invidia, et al. (2008). 'Human models of aging and longevity.' *Expert Opinion on Biological Therapy*, 8: pp. 1393–1405.

Atzmon, G., M. Rincon, et al. (2005). 'Biological evidence for inheritance of exceptional longevity.' *Mechanisms of Ageing and Development*, 126: pp. 341–45.

Slagboom, P.E., M. Beekman, et al. (2011). 'Genomics of

human longevity.' *Philosophical Transactions of the Royal Society: Biological Sciences*, 366: pp. 35–42.

Chapter Three
WHY WE AGE

Ageing is not necessary

Malthus' classic work is available from many publishers:

Malthus, R.T., (first published 1798). *An Essay on the Principle of Population.* Various publishers.

The essential message of this section is that our lives consist of a phase of programmed development followed by an unprogrammed phase of ageing. One very readable article on this is:

Austad, S.N. (2004). 'Is aging programed?' *Aging Cell*, 3: pp. 249–51.

The 'disposable soma'

The gerontologist Tom Kirkwood (Newcastle, UK), the spiritual father of the disposable soma theory, has written a very accessible book about his theory:

Kirkwood, T.B. (1999). *Time of Our Lives.* New York: Oxford University Press

Below are the details of the original publication and a more recent article, in which the theory is applied not only to the question of why we age, but also how:

Kirkwood, T.B. & R. Holliday, (1979). 'The evolution of ageing and longevity.' *Proceedings of the Royal Society: Biological Sciences*, 205: pp. 531–46.

Kirkwood, T.B. (2005). 'Understanding the odd science of aging.' *Cell*, 120: pp. 437–47.

The cost of sex

The unique study by biology professor Bas Zwaan (Wageningen, the Netherlands) was the first successful attempt to collect data by experimenting with fruit flies to prove the disposable soma theory:

Zwaan, B.J., R. Bijlsma, et al. (1995). 'Direct selection on lifespan in *Drosophila melanogaster*.' *Evolution*, 49: pp. 649–59.

The harmful consequences of sexual reproduction are divided between effects that are a *direct* result of mating and effects that are *indirect*, resulting from the fact that there is not one gender, but two, irrespective of whether the act of mating takes place or not. One example of the direct consequences of sexual reproduction in fruit flies is described in detail by the biology professor Linda Partridge (London, UK):

Cordts, R. & L. Partridge (1996). 'Courtship reduces longevity of male *Drosophila melanogaster*.' *Animal Behaviour*, 52: pp. 269–78.

An example of the consequences for hydras when they undergo sexual metamorphosis — the appearance of ageing — and the role played by stem cells in this phenomenon have been studied and described in detail:

Yoshida, K., T. Fujisawa, et al. (2006). 'Degeneration after sexual differentiation in hydra and its relevance to the evolution of aging.' *Gene*, 385: pp. 64–70.

Nishimiya-Fujisawa, C. & S. Kobayashi (2012). 'Germline stem cells and sex determination in Hydra.' *International Journal of Developmental Biology*, 56: pp. 499–508.

There is a great deal of debate among scientists over the correct explanation for the evolution of sexual reproduction. However, apart from all the argumentation and discussion, evolutionary experiments have also been carried out. They show a changing environment stimulates sexual reproduction, since the necessary investments then outweigh the costs:

Becks, L. & A.F. Agrawal (2010): 'Higher rates of sex evolve in spatially heterogeneous environments.' *Nature*, 468: pp. 89–92.

Aristocratic fruit flies

The following sources relate to the study by Tom Kirkwood and myself of the British nobility:

BBC News. 'Breed early, die young.' http://news.bbc.co.uk/2/hi/science/nature/241509.stm

Westendorp, R.G. & T.B. Kirkwood (1998). 'Human longevity at the cost of reproductive success.' *Nature*, 396: pp. 743–46.

Chapter Four
ASSESSORS OF THE FINITE

Insurance premium levels

The anecdote that forms the introduction to this discussion of 'life-expectancy tables' — also known as 'survival analysis' — is described in more historical detail in the 2010 lecture given by me on the occasion of the anniversary of the founding of Leiden University:

Westendorp, R.G. (2010) 'Passend of onaangepast? Over de menselijke levensloop in een snel veranderende omgeving.' ['Fit or unfit? On human life trajectories in a rapidly changing environment.'] (in Dutch). *Leiden University*. http://www.leidenuniv.nl/dies2010/dies_2010_oratie.pdf.

There is no consensus among scientists on how to interpret the phenomenon of proportional increase in mortality risks — that is, the Gompertz model. In the past, my colleagues and I have taken a stance on this: that there is a great difference between the doubling of a low risk of mortality and a doubling of a high risk. In the latter case, the additional number of people who die is much greater. This is not only numerically different, but the underlying biological explanations can vary widely. See:

Rozing, M.P. & R.G. Westendorp (2008). 'Parallel lines: nothing has changed?' *Aging Cell*, 7: pp. 924–27.

The impotence of prediction

Sickness and health are fundamentally unpredictable. Doctors are constantly (unpleasantly) faced with the fact that they are unable to make any precise predictions about the long-term development of a disease throughout an individual patient's life. The article below describes how Ancient Greek physician-philosophers recognised and dealt with this phenomenon:

Ierodiakonou, K. & J.P. Vandenbroucke (1993). 'Medicine as a stochastic art.' *The Lancet*, 341: pp. 542–43.

Chapter Five
SURVIVING IN HARSH CONDITIONS

An extraordinary find in chad

The references below indicate articles on the paleo-anthropological method, which uses fossil characteristics to make conclusions about the time when the first humans evolved. In this case, the question is when *Sahelanthropus* is likely to have lived. Current estimates place him 6 to 7 million years ago:

Zollikofer, C.P. & M.S. Ponce de León, et al. (2005). 'Virtual cranial reconstruction of *Sahelanthropus tchadensis*.' *Nature*, 434: pp. 755–59

An alternative method uses sensitive genetic analyses to estimate the time when the human lineage split from that of chimpanzees and bonobos. According to genetic researchers' latest calculations, that split began 5.5 million years ago, but was not definitively completed until 3 million years ago. These latest estimates are *not* in line with those produced by the paleo-anthropological method, and are the subject of debate:

Prüfer, K., K. Munch, et al. (2012). 'The bonobo genome compared with the chimpanzee and human genomes.' *Nature*, 486: pp. 527–31

When the life sciences and medicine are embedded in an evolutionary framework, the genetic basis of disease becomes easier to understand. It offers a logical explanation for why people become increasingly frail due to illness in old age. It replaces the prevailing understanding of organs as separate machines that are robbed of their functionality by specific biological processes. Evolutionary biology is an essential basis for achieving a better understanding of health and disease. See:

Nesse, R.M., C.T. Bergstrom, et al. (2009). 'Making evolutionary biology a basic science for medicine.' *Proceedings of the National Academy of Science of the USA*, 107: pp. 1800–07.

Taking the Leiden University debating tradition as a base, the physician Dr David van Bodegom and I have developed an appealing training method aimed at stimulating students to think in evolutionary terms. Its intention is to encourage them to develop a conceptual framework that will enable them to better understand the ageing process:

Bodegom, D. van, M. Hafkamp, et al. (2013). 'Using the master-apprentice relationship when teaching medical students academic skills: the Young Excellence Class.' *Medical Science Educator*, 23: pp. 80–83.

The book by Richard Dawkins mentioned in this chapter is:

Dawkins, R. (2006) *The Selfish Gene — 30th anniversary edition*. Oxford, New York: Oxford University Press.

The gold coast of africa

Between 2002 and 2012, we at the Department of Gerontology and Geriatrics at the Leiden University Medical Centre (LUMC) carried out research in the Garu District of north-eastern Ghana, close to the Togolese border. The study was initiated by the anthropologist Dr Hans Meij, and was later continued by the physician Dr David van Bodegom. Both men wrote their doctoral theses on this research. A whole host of (master's) students has followed in their footsteps. For the relationship between socio-economic class and mortality, see:

Bodegom, D. van, L. May, et al. (2009). 'Socio-economic status is highly correlated with mortality risks in rural Africa.' *Transactions of the Society of Tropical Medicine & Hygiene*, 103: pp. 795–800.

Taking the human life trajectory — from babyhood to death — as his basic structure, the epidemiologist Professor Tim Spector (London, UK) describes the (evolutionary) hereditary background to personality, physical characteristics, risk of illness, sex, and risk-taking, using case studies of identical and fraternal twins:

Spector, T. (2012). *Identically Different: Why You Can Change Your Genes.* London: Weidenfeld & Nicolson.

Resistance to infectious disease

When humans began to work the land and keep livestock, many different viruses, bacteria and parasites were presented with new opportunities, resulting in the spread of all kinds of diseases and scourges. This was accompanied by an increase in the number of animal-to-human infections. As human communities grew in size, and the interactions between (groups of) humans became increasingly intensive, diseases that spread without direct contact, such as smallpox, were also able to advance:

Stearns, S.C. & J.C. Koella (Eds.) (2008) *Evolution in Health and Disease: Second Edition*. Oxford: Oxford University Press

Earlier, the Leiden-based rheumatologist Professor Tom Huizinga and I mimicked human immune reactions in blood samples. This led to ideas about why women who live to a great age have less chance of a successful pregnancy. The article below describes in more detail how the innate immune system increases resistance to infections, while simultaneously increasing the chance that a pregnancy will end preterm in a spontaneous abortion:

Bodegom, D. van, L. May, et al. (2007). 'Regulation of human life histories: the role of the inflammatory host response.' *Annals of the New York Academy of Sciences*, 1100: pp. 84–97.

We were able to demonstrate the phenomenon known to biologists as the 'quality-quantity trade-off' among humans:

Meij, L.L., D. van Bodegom, et al. (2009). 'Quality-quantity trade-off of human offspring under adverse environmental conditions.' *Journal of Evolutionary Biology*, 22: pp. 1014–23.

I recommend two references for those who would like to know more about the idea of balancing selection for immune defences. First of all, my colleague Tom Kirkwood and I developed a mathematical model to describe this. Later, we were able to demonstrate the suspected natural-selection mechanism at work on the immune system of subjects in our research area in Ghana:

Drenos, F., R.G. Westendorp, et al. (2006). 'Trade-off mediated effects on the genetics of human survival caused by increasingly benign living conditions.' *Biogerontology*, 7: pp. 287–95.

Kuningas, M., L. May, et al. (2009). 'Selection for genetic variation inducing pro-inflammatory responses under adverse environmental conditions in a Ghanaian population.' *PLoS One*, 4: e7795.

The benefit of grandmothers

I recommend two articles for those who want to explore this broad area of research in more depth. The first concerns the work of the group led by the sociologist Dr Fleur Thomese (Amsterdam), which made use of data from three generations in Dutch families. The second is the work of my own research group on the impact of (grand)parents on the number of children born, and their survival, in Ghana. Both studies were part of a joint research programme financed by the Netherlands Organisation for Scientific Research (NWO) entitled 'Aging Societies: Human Victory or Evolutionary Trap?':

Thomese, F. & A.C. Liefbroer (2013). 'Child care and child births: the role of grandparents in the Netherlands.' *Journal of Marriage and Family*, 75: pp. 403–21.

Bodegom, D. van, M. Rozing, et al. (2010). 'When grandmothers matter.' *Gerontology*, 56: pp. 214–16.

Chapter Six
OUR INCREASED LIFE EXPECTANCY

What we used to die of

For a nice overview of John Snow's original work and the impact it had on the theory and practice of medicine, see:

Fine, P., C.G. Victora, et al. (2013). 'John Snow's legacy: epidemiology without borders.' *The Lancet*, 381: pp. 1302–11.

The first description of epidemiological transition is discussed in:

Omran, A.R. (reprint, first published in 1971) (2005) 'The epidemiologic transition: a theory of the epidemiology of population change.' *The Milbank Quarterly*, 83: pp. 731–57

On the decline of violence in the world, see:

Pinker, S. (2011). *The Better Angels of Our Nature: a history of violence and humanity*. London: Penguin Books.

The new killers

A good recent study of the causes of death of mummified humans, showing that cardiovascular disease also occurred in antiquity, is:

Thompson, R.C., A.H. Allam, et al. (2013). 'Atherosclerosis across 4000 years of human history: the Horus study of four ancient populations.' *The Lancet*, 381: pp. 1211–22.

More than three-quarters of the world's cases of chronic degenerative disease occur in low-income or middle-income countries. Risk factors that are associated with affluence, such as high blood pressure, high cholesterol levels, and too little exercise play an ever-increasing role in the pattern of disease in such countries. See:

World Health Organization (2009). *Global Health Risks: mortality and burden of disease attributable to selected major risks.* Geneva: WHO Press.

In order to better understand the explosion of diabetes in cities — urban diabetes — public and private partners have joined forces and are currently mapping the urban-diabetes challenges in a number of cities across the world. The aim is to generate a body of collective knowledge about what is working today, where the challenges are, and what the priorities should be for the future:

http://citieschangingdiabetes.com/

A revolution in medical technology

For an overview of everything possible in the area of cardiovascular disease, I recommend the following websites:

http://www.heartfoundation.org.au

http://www.bhf.org.uk

http://www.heart.org.

On the early signs of atherosclerosis among fallen American servicemen:

Webber, B.J., P.G. Seguin, et al. (2012) 'Prevalence of and risk factors for autopsy-determined atherosclerosis among US service members, 2001–2011.' *Journal of the American Medical Association*, 308: pp. 2577–83.

An extra weekend every week

The internationally renowned Max Planck Institute for Demographic Research (MPIDR, Rostock, Germany), under the leadership of the demographer Professor Jim Vaupel, carries out research on our increasing life expectancy, also from an evolutionary biology perspective. Jim Vaupel has developed a realistic scenario for the permanent increase in life expectancy. For an analysis and a projection of the increase in our life expectancy, see:

Oeppen, J. & J.W. Vaupel (2002). 'Demography. Broken limits to life expectancy.' *Science*, 296: pp. 1029–31.

Vaupel, J. W. (2010). 'Biodemography of human aging.' *Nature*. 464: pp. 536–42

The extrapolation of increasing life expectancy caused a veritable storm of public controversy. For instance, see:

http://www.theatlantic.com/features/archive/2014/09/why-i-hope-to-die-at-75/379329/

As another example of this relentless rise, the Australian government's 2015 Intergenerational Report projects that in 2054–55, life expectancy for Australians at birth will be 95.1 years for men and 96.6 years for women. (In 2015, the figures were 91.5 years for men and 93.6 for women.) The report also predicts that by 2054–55 nearly 2 million Australians will be aged 85 and over, of which approximately 40,000 will be over 100 years old.

To read more of this report, go to: http://www.treasury.gov.au/PublicationsAndMedia/Publications/2015/2015-Intergenerational-Report

Chapter Seven
LEGIONS OF THE OLD

The gravedigger

Data on age at death from the nineteenth century onwards can be found on the website of Statistics Netherlands. An alternative source of information is the Human Mortality Database, which is maintained by the Max Planck Institute for Demographic Research (MPIDR). A graphic representation of the number of deaths per age group over time can be found on the website of the Leyden Academy on Vitality and Ageing. The gravedigger's story is based on this data. For graphic representations, see:

http://www.leydenacademy.nl/Ageing/What_can_be_learned_from_gravediggers

The Lochem gravedigger Lenderink was a real person. His annual report from 1889 is part of Dutch cultural history:

Spruit, R. (1986). *De dood onder ogen: een cultuurgeschiedenis van sterven, begraven, cremeren en rouw* [*Facing Death: a cultural history of death, burial, cremation and mourning*]. (in Dutch) Houten: De Haan.

From pyramid to skyscraper

For a theoretical consideration of democratic transition — the change in the structure of the population due to a shift in the causes of death — see:

Galor, O. & D.N. Weil (2002). 'Population, technology, and growth: from Malthusian stagnation to the demographic transition and beyond.' *The American Economic Review*, 90: pp. 806–28.

For the concept of the second demographic transition — the change in fertility patterns due to reduced mortality — see:

Kaa, D.J. van de (1987) 'Europe's second demographic transition.' *Population Bulletin*, 42: pp. 1–59.

For a description of the demographic transition as observed by my research group in our study area in Ghana, see:

Meij, L.L., A.J. de Craen, et al. (2009). 'Low-cost interventions accelerate epidemiological transition in Upper East Ghana.' *Transactions of the Royal Society of Tropical Medicine and Hygiene*, 103: pp. 173–78.

Young and old-age dependency ratios

For more on the development of demographic ratios in the Netherlands, I recommend the websites of Statistics Netherlands (the Dutch national statistics office) and the Netherlands Interdisciplinary Demographic Institute (NIDI). For data on young- and old-age dependency ratios for countries in the world — past, present, and future trajectories — see population statistics of the United Nations at:

http://esa.un.org/unpd/ppp/Figures-Output/Population/PPP_Total-Dependency-Ratio.htm

Please note that, in this book, I have followed the age categories of Statistics Netherlands, which employs a higher cut-off for defining young people (below 20 years) than the United Nations (below 15 years). This explains the disparities with the numbers on the UN site (the young-age dependency ratios are higher in my text). In a same vein, some statistical offices use a cut-off of 60 years to define older people, which has the effect of increasing the old-age dependency ratios even further. Decisions on these cut-off points have considerable socio-economic impact, and are thus the subject of intense political debate. Irrespective of the specific cut-offs, the trends over time will not be different.

A fantastic tool for creating your own analyses, graphics, and diagrams can be found at:

Gapminder: http://www.gapminder.org

For the costs of raising kids in Australia, see:

NATSEM at the University of Canberra. The cost of raising children in Australia. AMP.NATSEM Income and Wealth Report issue 33, May 2013.

Chapter Eight
AGEING IS A DISEASE

What causes cancer?

For an overview of the relation between ageing and cancer, I recommend the work of the molecular biologist Professor Jan Hoeijmakers (Rotterdam), a researcher of international repute in the field of DNA damage, the pathogenesis of cancer, and ageing:

Hoeijmakers, J.H. (2009). 'DNA damage, aging, and cancer.' *New England Journal of Medicine*, 361: pp. 1475–85.

The occurrence of random mutations that arise during DNA replication in normal, non-cancerous stem cells is related to the risk of cancer. The difference in stem-cell divisions may explain the huge differences in the various types of cancer. This could also explain the element of 'bad luck' — that is, the impotence of prediction. See:

Tomasetti, C. & B. Vogelstein (2015) 'Variation in cancer risk among tissues can be explained by the number of stem cell divisions,' *Science*, 347: pp. 78–81.

Normal ageing does not exist

There is no difference between ageing and the development of disease in old age. I demonstrate this in two stages. The first stage is based on the theory of the origin of disease developed by the epidemiologist Professor Kenneth Rothman (Boston, USA).

Disease does not have just *one* cause, but always results from the interplay of partial causes, which themselves are not sufficient to explain the genesis of disease alone. In this book, I explain this using the example of the Concorde disaster. Rothman's original 1976 publication in the *American Journal of Epidemiology* is the best, in my opinion, because it presents the facts in a plain and simple way. This model of pathogenesis has been revised and adapted many times. For the most recent version, see:

Rothman, K.J. & S. Greenland (2004). 'Causation and causal inference in epidemiology.' *American Journal of Public Health*, 95: pp. 144–50.

In the second stage of my argument, I apply the theory of partial causes of disease to the principle of ageing as explained by the accumulation of small forms of damage that come together to explain a disease or impairment. I have tried to explain this in this book using the game of bingo as an analogy. See:

Izaks, G.J. & R.G. Westendorp (2003). 'Ill or just old? Towards a conceptual framework for the relation between ageing and disease.' *BMC Geriatrics*, 3: e7.

Wensink, M., R.G. Westendorp, et al (2014). 'The causal pie model: an epidemiological method applied to evolutionary biology and ecology.' *Ecology and Evolution*, 4:1924–30.

The dementia epidemic

For a reference to the stirring neuropathological study that has been dubbed 'The Nun Study', see:

Snowdon, D. *Aging with Grace* (2001). New York: Bantam Books.

Predicting the number of dementia patients in the near and distant future is part of a public debate on the subject. I take the view that the end of the dementia epidemic is in sight. I base this opinion on the knowledge that dementia in old people has a multi-causal explanation and that its various partial causes can be influenced effectively. See:

Savva, G.M., S.B. Wharton, et al. (2009). 'Age, neuropathology, and dementia.' *The New England Journal of Medicine*, 360: pp. 2302–09.

Since research shows that cardiovascular disease plays a large role in the cause of dementia — and the prevalence of cardiovascular disease has fallen drastically in the past — the risk of developing dementia *must* have shrunk in the past ten to twenty years. Various studies were published in the course of 2012–2013 that include figures to support this interpretation:

Schrijvers, E.M., B.F. Verhaaren, et al. (2012). 'Is dementia incidence declining? Trends in dementia incidence since 1990 in the Rotterdam Study.' *Neurology*, 78: pp. 1456–63.

Qiu, C., E. von Strauss, et al. (2013). Twenty-year changes in dementia occurrence suggest decreasing incidence in central Stockholm, Sweden.' *Neurology*, 80: pp. 1888–94.

Christensen, K., M. Thinggaard, et al. (2013). 'Physical and cognitive functioning in people older than 90 years: a comparison of two Danish cohorts born 10 years apart.' *The Lancet*, 382: pp. 1507–13.

Matthews, F.F., A. Arthur, et al. (2013). 'A two-decade comparison of prevalence of dementia in individuals aged 65 years and older from three geographical areas of England: results of the cognitive function and ageing study I and II.' *The Lancet*, 382: pp. 1405–12.

Frailty

Although the principle of frailty is easy to define, it is difficult to estimate an individual person's degree of frailty, just as the course of a disease or the event of death is difficult to predict. This difficulty is reflected in the fact that so many definitions of frailty have been put forward. The geriatrician Professor Marcel Olde Rikkert (Nijmegen, the Netherlands) has shown before that

— disturbingly — some people are sometimes declared to be frail and sometimes not, depending on the definition used. There is currently no gold standard:

Iersel, M.B. van, & M.G. Rikkert (2006). 'Frailty criteria give heterogeneous results when applied in clinical practice.' *Journal of the American Geriatric Society* 54: pp. 728–29.

It is not possible to predict frailty, illness, and death for individuals, but it can be done for groups. A simple and decisive measure is walking speed:

Studenski, S., S. Perera, et al. (2011). 'Gait speed and survival in older adults.' *Journal of the American Medical Association*, 305: pp. 50–58.

Chapter Nine
THE BIOLOGY OF AGEING

For more background information on the biology of ageing, I recommend the website of the National Institute of Aging (Bethesda, USA)

http://www.nia.nih.gov/health/publication/biology-aging.

For those who would like to read more about proximate and ultimate causation, I recommend the work of the Ukrainian-American geneticist Theodosius Dobzhansky (1900–1975). He was one of the most important proponents of 'modern evolutionary synthesis', which unites genetics — in particular, Mendel's laws — with evolutionary theory.

Dobzhansky, T. (1973). 'Nothing in biology makes sense except in the light of evolution.' *The American Biology Teacher*, 35: pp. 125–29.

As part of an experiment for his master's students, the behavioural biologist Professor Carel ten Cate (Leiden) replicated the fieldwork of the Dutch behavioural biologist Niko

Tinbergen (1907–1988):

Cate, C. ten, W. Bruins, et al. (2009). 'Tinbergen revisited: a replication and extension of experiments on the beak colour preferences of herring gull chicks.' *Animal Behaviour*, 77: pp. 795–802.

Accelerated ageing

One of the most eminent scientists who put 'progeroid syndromes' on the map is the pathologist Professor George Martin (Washington, USA). In the article recommended below, he discusses the various syndromes, and indicates how they differ from the usual human ageing process:

Martin, G.M. (2005). 'Genetic modulation of senescent phenotypes in *Homo sapiens*.' *Cell*, 120: pp. 523–32.

Oxygen radicals

Professor Gems (London, UK), a molecular biologist, has carried out some unique experiments in which he used genetic manipulation to switch off the natural antioxidant mechanisms gradually in nematode worms and fruit flies. This had *no* impact on their survival rate. For a critical discussion of the 'oxygen radical theory of ageing', see:

Gems, D. & R. Doonen (2009). 'Antioxidant defence and aging in *C. elegans*: is the oxidative damage theory of aging wrong?' *Cell Cycle*, 8: pp. 1681–87.

Researchers have gathered data on a total of 296,707 people who took part in various experimental studies on the effect of supplements containing antioxidants. The conclusion is that such supplements have no beneficial effect:

Bjelakovic, G., D. Nikolova, et al. (2012). 'Antioxidant supplements for prevention of mortality in healthy participants and patients with various diseases.' *The Cochrane Library*, 14 March.

Insulin and growth hormone

The idea that humans have evolved to be economical with energy — known as the 'thrifty gene hypothesis' — was first developed by the American geneticist James Neel (1915–2000), who wanted to find a plausible explanation for the increasing prevalence of diabetes in our modern environment. This theory may also explain why people quickly become obese during times of abundance. The following article deals with the original theory:

Neel, J.V. (1962). 'Diabetes mellitus: a 'thrifty' genotype rendered detrimental by "progress"?' *The American Journal of Human Genetics*, 14: pp. 353–62.

The theory is controversial. Some scientists have formulated opposing theories, in which they argue that there was selective pressure *against* obesity in our original environment, since an athletic constitution was advantageous in an environment where humans could be preyed on by other species:

Speakman, J.R. (2007). 'A nonadaptive scenario explaining the genetic predisposition to obesity: the "predation release" hypothesis.' *Cell Metabolism*, 6: pp. 5–12.

Researchers believe that the activity of genes is adapted for optimal metabolism in the circumstance in which an individual finds itself. Thus, circumstances in the womb should be able to induce life-long changes in an individual's genetic material. This is called 'epigenetics'. For example, it is thought that the metabolism of children carried during the Dutch Winter Famine of 1944 is more 'thrifty' than others'. The epidemiologist Dr Tessa Rosenboom (Amsterdam) has written a wonderful book on this subject:

Rosenboom, T. & R. van de Krol (2010). *Baby's van de Hongerwinter: de onvermoede erfenis van ondervoeding* [*Babies of the Winter Famine: the unexpected legacy of malnutrition*] (in Dutch). Amsterdam: Augustus.

Professor David Barker (Southampton, UK) is the original advocate of the theory that may explain why children of the Winter Famine have higher incidences of obesity and cardiovascular disease. See:

Hales, C.N. & D.J. Barker (1992). 'Type 2 (non-insulin-dependent) diabetes mellitus: the thrifty phenotype hypothesis.' *Diabetologia*, 35: pp. 595–601.

The biologist Professor Eline Slagboom (Leiden) and her colleagues were the first to show that the genes of those who survived the Winter Famine are calibrated differently to their siblings who did not survive:

Heijmans, B.T., E.W. Tobi, et al. (2008). 'Persistent epigenetic differences associated with prenatal exposure to famine in humans.' *Proceedings of the National Academy of Sciences of the United States of America*, 105: pp. 17046–17049.

The following are two articles on the Leiden 85-plus Study and the Leiden Longevity Study, in which we tested whether variations in the insulin/IGF-1 signalling pathway can explain differences in longevity:

Heemst, D van, M. Beekman, et al. (2005). 'Reduced insulin/IGF-1 signaling and human longevity.' *Aging Cell*, 4: pp. 79–85.

Rozing, M.P., R.G. Westendorp, et al. (2009). 'Human insulin/IGF-1 and familial longevity at middle age.' *Aging (Albany, NY)*, 1: pp. 714–722.

The article recommended below is an overview for advanced readers which covers the connection between insulin/IGF-1 signalling pathway, environment, selection, adaptation, growth, metabolism, and lifespan:

Gems, D. & L. Partridge (2013). 'Genetics of longevity in model organism: debates and paradigm shifts.' *Annual Review of Physiology*, 75: pp. 621–644.

Should we eat less?

Many books have been written about 'caloric restriction', and entire movements have been based on the theory. Caloric restriction is erroneously seen as the Holy Grail that can slow the pace of ageing under any circumstances. In mice, its life-prolonging effect appears to be strongly dependent on the animals' genetic background, and it is not possible to produce a beneficial effect in all species and in all individual members of a species:

Liao, C.Y., B.A. Rikke, et al. (2010). 'Genetic variation in the murine lifespan response to dietary restriction: from life extension to life shortening.' *Aging Cell*, 9: pp. 92–95.

The next question is whether caloric restriction works in humans. It is not easy to find this out by experimental means. That would require subjects who are willing to subject themselves to a normal or restricted caloric intake for their entire lives. Nevertheless, preparations are underway for a long-term experiment of this kind. For those who are interested:

Caloric restriction in humans: http://calerie.dcri.duke.edu/

Finally, I recommend two recent publications concerning two long-term experiments with caloric restriction in rhesus monkeys in the United States. The preliminary conclusion is that some obesity-related diseases, such as diabetes, do not occur in the calorie-restricted group. A beneficial influence on lifespan has not (yet) been demonstrated:

Colman, R.J., R.M. Anderson, et al. (2009). 'Caloric restriction delays disease onset and mortality in rhesus monkeys.' *Science*, 325: pp. 201–204.

Mattison, J.A., G.S. Roth, et al. (2012). 'Impact of caloric restriction on health and survival in rhesus monkeys from the NIA study.' *Nature*, 489: pp. 318–321.

Chapter Ten
LONG MAY WE LIVE

More years of illness

For a definition of and figures on healthy life expectancy, I recommend the National Public Health Compass provided by the Dutch National Institute for Public Health and the Environment, as well as the Dutch national statistical bureau, Statistics Netherlands. Trends in (healthy) life expectancy in the Netherlands over the past 30 years are described in:

Engelaer, F.M., D. van Bodegom, et al. (2013). 'Sex differences in healthy life expectancy in the Netherlands.' *Annual Review of Gerontology and Geriatrics*, 33: pp. 361–371 (11).

More years without impairment

Screening and preventive measures are often introduced without knowing whether any resultant intervention will lead to a predetermined health effect. Often, we lack the tools to identify the right individuals, or no effective intervention is available anyway. In the latter case, there is simply no reason for screening people. The expert panel mentioned in the text, which systematically evaluated the arguments for and against screening, published an article on its assessments:

Drewes, Y.M., J. Gussekloo, et al. (2012). 'Assessment of appropriateness of screening community-dwelling older people to prevent functional decline. *Journal of the American Geriatric Society*, 60: pp. 42–50.

There is controversy surrounding the lowering of the upper age-limit for breast-cancer screening, although it was the result of careful consideration on the part of the Health Council of the Netherlands: 'A model calculation by the National Evaluation Team for Breast Cancer Screening (LETB) determined that the beneficial effects of screening for breast cancer outweigh the negative effects, up to the age of 75.' On the basis of this, a court in The Hague ruled on 9 February 2010 that an age limit of 75 can be applied in breast-cancer screening. The ruling concerned an appeal against the Dutch State brought by three women, the Clara Wichmann Test Case Fund Foundation, and the Netherlands Breast Cancer Association.

Gezondheidsraad: Commissie WBO (2001). *Wet bevolkingsonderzoek: landelijke borstkanker-screening (2).* [Health Council of the Netherlands: Commission on the Population Screening Act (WBO) (2001). *Population Screening Act: national breast cancer screening (2).*] (in Dutch). The Hague: publication no. 2001/2.

The ragged end

Ideas about healthy ageing go back to the (false) claim by the American doctor James Fries that in the future we will be able to remain healthy to the last moment — thanks to our ability to eliminate disease — and then die between the ages of 70 and 90 due to the mechanism of ageing:

Fries, J.F. (1980). 'Aging, natural death, and the compression of morbidity.' *The New England Journal of Medicine*, 303: pp. 130.135.

The demographer Professor John Wilmoth (Berkeley, USA) uses data from Sweden and Japan from the past fifty years to demonstrate that this simply does not happen. In doing so, he also shows that it is never likely to happen, because it is not only the age at death that is rising, but also the maximum age reached. This means that all deaths are shifted to later in life, and that is precisely what was demonstrated in the genetic experiments carried out with nematode worms:

Wilmoth, J.R., & S. Horiuchi (1999). 'Rectangularization revisited: variability of age at death within human populations.' *Demography*, 36: pp. 475–495.

Wilmoth, J.R., L.J. Deegan, et al. (2000). 'Increase of maximum life-span in Sweden, 1861–1999.' *Science*, 289: pp. 2366–2368.

Kirkwood, T.B., & C.E. Finch (2002). 'Ageing: the old worm turns more slowly.' *Nature*, 419: pp. 794–795.

Chapter Eleven
THE QUALITY OF OUR EXISTENCE

What is healthy?

For more information on the World Health Organization (WHO) definition of health, see: http://www.who.int/about/definition/en.

For the 'disability paradox', see:

Albrecht, G.L. & P.J. Devlieger (1999). 'The disability paradox: high quality of life against all odds.' *Social Sciences & Medicine*, 48: pp. 977–988.

Convinsky, K.E., A.W. Wu, et al. (1999). 'Health status versus quality of life in older patients: does the distinction matter?' *The American Journal of Medicine*, 106: pp. 435–440.

The Leiden 85-plus study

In 1997, Annetje Bootsma, Eric van Exel, Margaret von Faber, Jacobijn Gussekloo, Dr Gooke Lagaay, the late Professor Dick Knook, and I took a random sample of the people of Leiden. All 599 85-year-olds were visited over a period of two years and monitored over a period of more than ten years for health, illness, quality of life, and death. This interdisciplinary cooperation between doctors and gerontologists from Leiden and the Amsterdam-based cultural anthropologists the late Dr Els van Dongen and Professor Sjaak van der Geest delivered a number of new and enriching insights. The study into being healthy and feeling healthy as described in this book is a core publication from that period:

Faber, M. von, A. Bootsma-van der Wiel, et al. (2001). 'Successful aging in the oldest old: who can be characterized as successfully aged?' *JAMA International Medicine*, 161: pp. 2694–2700.

In this partial study, two classic, mutually exclusive interpretations of successful ageing were tested among 85-year-olds. For the original articles in which these theories were presented, see:

Rowe, J.W. & R.L. Kahn (1987). 'Human aging: usual and successful.' *Science*. 237: pp. 143–149.

Baltes, P.B. & B.M. Baltes (1990). 'Psychological perspectives on successful aging: the model of selective optimization with compensation.' In: Baltes, P.B & B.M. Baltes (Eds.) (1990). *Successful Ageing: Perspectives from the Behavioural Sciences*, pp. 1–34. Cambridge: The Press Syndicate of the University of Cambridge.

A rating for life

The sociologist Professor Ruut Veenhoven (Rotterdam) investigates the social conditions that contribute to human happiness. He does this by asking people *one* single question about their lives: 'How satisfied are you, all things considered, with the life you lead?' The basis for his research is described here:

Veenhoven, R. (2000). 'The four qualities of life.' In: McGillivray, M. & M. Clarke (Eds.) (2006). *Understanding Human Well-Being*, pp. 74–100. New York: United Nations University Press.

Measuring feelings can be very subjective, but is nonetheless a useful complement to more objective data when comparing quality of life across countries. Subjective data can provide a personal evaluation of an individual's health, education, income, personal fulfilment, and social conditions. For a more elaborate discussion of the topic, see:

WHO (2013) *World Happiness Report*. New York, WHO.

For the report of the Commission on Wellbeing and Policy, see:

O'Donnell, G., Deaton, A., et al. (2014). *Wellbeing and Policy.* London: Legatum Institute.

When asked to rate their general satisfaction with life on a scale from 0 to 10, people across the OECD gave it a 6.6 grade. Life satisfaction is not evenly shared across the OECD, however. See:

http://www.oecdbetterlifeindex.org/topics/life-satisfaction/

Life satisfaction levels fall only in the final year of life, which is unsurprising in view of the fact that death is usually preceded by increasing infirmity. This effect has been demonstrated in follow-up studies in several developed countries:

Gerstorf, D., M. Hidajat, et al. (2010). 'Late-life decline in well-being across adulthood in Germany, the United Kingdom, and the United States: something is seriously wrong at the end of life.' *Psychology and Aging*, 25: pp. 477–85.

Chapter Twelve
VITALITY

The new 'Ages of Man and Woman'

As a concept, the new 'Ages of Man and Woman' consist of stages of prevention, multimorbidity, frailty, and dependency. My colleague Marieke van der Waal and I first published this idea in:

Westendorp, R.G. & M. van der Waal (2011). 'Anders kijken naar de ouderenzorg.' ['Taking a different view of old-age care.'] (in Dutch). *Zorgmarkt*, 11: pp. 13–16.

The staggering finding that we in the West fail to treat, or treat adequately, three-quarters of cases of high blood pressure — an important risk factor for dementia in old age — can be found in:

Chow, C.K., K.K. Teo, et al. (2013). 'Prevalence, awareness, treatment and control of hypertension in rural and urban communities in high-, middle-, and low-income countries.' *Journal of the American Medical Association*, 310: pp. 959–68.

Support for the statement 'it is better to smoke than have a small social network' can be found in:

Holt-Lunstad, J., T.B. Smith, et al. (2010). 'Social relationships and mortality risks: a meta-analytic review.' *PLoS Medicine* 7, e1000316.

The present design of the healthcare system is associated with major problems in the treatment of various currently occurring complaints. An initial orientation in this subject matter can be found in:

Boyd, C.M., J. Darer, et al. (2005). 'Clinical practice guidelines and quality of care for older patients with multiple comorbid diseases: implications for pay and performance.' *Journal of the American Medical Association*, 294: pp. 716–24.

More than 60 per cent of doctors in the Netherlands believe that seriously ill patients in the final stages of life are treated for longer than desirable; 22 per cent disagree. The rest indicated no opinion.

Visser, J. (2012). 'De arts staat in behandelmodus.' ['The doctor is in treatment mode'] (in Dutch). *Medisch Contact*, 22: pp. 1326–29.

The view is vocally expressed by the US-based surgeon and popular author Dr Atul Gawande. See:

Gawande, A. (2014). *Being Mortal: illness, medicine and what matters in the end*. London: Profile Books.

In this book, I argue that professionals in long-term care are often caught in a medical, technical, and legal way of thinking, while not listening enough to the people they are treating. As justification of this bold statement, I present here the example of under-nutrition.

Under-nutrition is regarded as a medial alarm signal in the care of old people. For this reason, the various professions have developed guidelines to identify and treat under-nutrition in old people. The incidence of under-nutrition is also seen as a quality indicator of (medical) care. For this reason, institutions work according to protocols, and public-health inspections check them for compliance. All these actions are highly remarkable, since the Health Council of the Netherlands has determined that under-nutrition cannot be unambiguously defined, its cause cannot generally be well identified, and general guidelines to identify and treat it cannot be drawn up:

Health Council of the Netherlands (2011). *Undernutrition in the Elderly*. The Hague, Health Council of the Netherlands 2011/32E

Thus, under-nutrition is an example of my claim that in nursing and care homes we often miss the point, and professionals are too rarely in a position to provide for the wishes of patients. Under-nourishment needs to be approached in a completely different way. Wija van Staveren (Wageningen, the Netherlands), Professor of Nutrition and the Elderly, has shown that setting the table, serving food in tureens, and allowing residents time to eat at their own pace can turn the trend of weight loss in institutions into a trend of weight gain:

Nijs, K.A., C. de Graaf, et al. (2006). 'Effect of family-style mealtimes on quality of life, physical performance, and body weight of nursing home residents: cluster randomised controlled trial.' *British Medical Journal*, 322: pp. 1180–84.

Optimism and zest for life

Social scientists have long stressed that vitality plays an important part in wellbeing in old age. Vitality is characterised by features such as introspection, positive emotions, energy, involvement, resilience, self-confidence, independence, and a sense of purpose. See:

Ryan, R.M. & C. Frederick (1997). 'On energy, personality, and health: subjective vitality as a dynamic reflection of well-being.' *Journal of Personality*, 65: pp. 529–65.

In the context of ageing, the Leyden Academy on Vitality and Ageing has created an operating definition of vitality: the ability of a person to set ambitions that are appropriate for their life situation, and to realise those goals despite functional limitations. The importance of achieving individual goals for a sense of wellbeing has often been stressed and is essential in old age when ambitions and goals are seen as less than obvious:

Westendorp, R.G., B. Mulder, et al. 'When vitality meets longevity. New strategies for health in later life.' In: Kirkwood, T.B. & C.L. Cooper (Eds.) (2013). *Wellbeing in Later Life: a complete reference guide, Volume IV: Wellbeing in Later Life*. London: John Wiley & Sons, Ltd.

For a reference to the revolution in thinking that took place in cultural anthropology, see:

Marcus, L. & A. Marcus (1988). 'From soma to psyche: the crucial connection. Part 2. Cross-cultural medicine, decoded: learning about "us" in the act of learning about "them". *Family Medicine*, 20: pp. 449–57.

Grey is not black and white

Medical Delta, a partnership between the universities of Leiden, Delft, and Rotterdam and the South-Holland provincial government, applies its members' knowledge and expertise to finding solutions to the social and personal consequences of the ageing society. These activities are bundled in the VITALITY! Programme. The research project, *Many Shades of Grey: ambitions of 55+*, paints a picture of the views, desires, and needs of old people as they seen by themselves. The questions and answers are presented from a representative sample of 650 people aged 55 or older. The questions were developed in a preparatory study with panels of older people. See: http://www. leydenacademy.nl/UserFiles/file/Report_Shades _of_grey_LR.pdf.

Chapter Thirteen
THE NEW LIFE TRAJECTORY

75 Is the new 65

In his farewell lecture, the sociologist Professor Kees Knipscheer (Amsterdam) argued for over-50s to begin thinking of ending their first career and starting a second one. The second career would ideally be more focussed on acquired expertise, and should not be allowed to degenerate into a form of early retirement. A second career can be more flexible, whether it involves part-time working or not. Over-50s may work as self-employed entrepreneurs or as employees with a variable pay system. The social-security system should be integrated with the already-initiated changes to the laws regulating retirement and pre-retirement:

Knipscheer, K. (2005). *De uitdaging van de tweede adolescentie.* [*The Challenge of the Second Adolescence.*] (in Dutch). Amsterdam: Oratie Vrije Universiteit.

Who is responsible for what?

The original version of Martina Rosenberg's book *Mother, When*

Will You Finally Die?:

Rosenberg, M. (2013) *Mutter, wann stirbst du endlich?* (in German). Munich. Blanvalet Verlag.

A birthday rhyme

Figures concerning remaining life expectancy can be found on the website of Statistics Netherlands. For an interpretation of those statistics, see:

Hintum, M. van (2013). 'Nog nooit zo lang gezond'. ['Healthy Longer than Ever'] (in Dutch). *De Volkskrant*, 2 March 2013.

A PRESCRIPTION FOR THE FUTURE

My interpretation of current dietary advice does not differ significantly from that of Professor Walter Willett (Harvard, USA), the world's most-cited professor of nutrition, as they are presented in his bestseller:

Willett, W.C. (2005). *Eat, Drink, and Be Healthy: the Harvard Medical School guide to healthy eating*. New York: Free Press.

John Ioannides, Professor of Epidemiology (Stanford, USA), is even more direct, pointing out that research results into eating habits are often 'too good to be true'.

Ioannides, J.P. (2013). 'Implausible results in human nutrition research.' *BMJ*, 347: p. 6698

One Dutch advocate of behavioural change as a guiding principle for new public health is Willem van Mechelen (Amsterdam, the Netherlands), Professor of Public and Occupational Health. He has written on the subject, in collaboration with colleagues, in:

Matheson, G.O., M. Klügl, et al. (2013). 'Prevention and Management of Non-Communicable Disease: the IOC Consensus Statement, Lausanne 2013.' *British Journal of Sports Medicine*, 47: pp. 1003–11.

Increasing frail older people's own strength, listening to them, and introducing professional action on the basis of what they say is the central principle of the Dutch National Care for the Elderly Programme:

http://www.nationaalprogrammaouderenzorg.nl/english/the-national-care-for-the-elderly-programme/

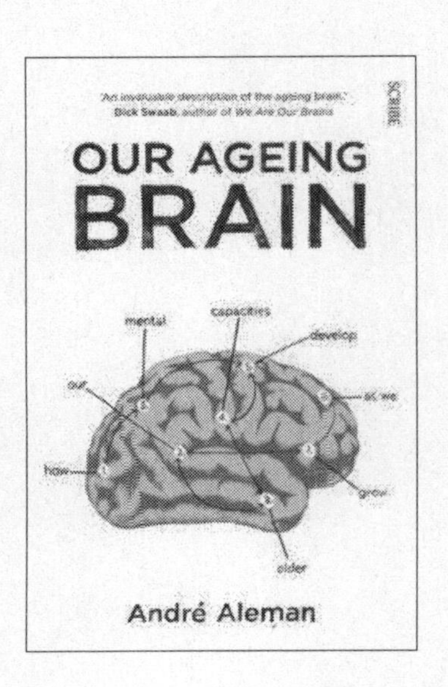

Our Ageing Brain
ANDRÉ ALEMAN

An international bestseller delivering good news on brain function and ageing

We all worry sometimes that our brains — particularly our memories — just don't work as well as they used to. In this illuminating book, internationally acclaimed Dutch neuroscientist André Aleman shows that although the decline in our mental capacities begins earlier than we think, this is not such a bad thing. In fact, older people are more resistant to the effects of stress, cope better with their emotions and with complex situations, and are — generally speaking — happier than their younger counterparts.

'A book you can't put down, no matter how old you are.' *Psychologie*